RAIDERS

GREAT EXPLOITS OF THE SECOND WORLD WAR

JOHN LAFFIN

SUTTON PUBLISHING

This book was first published in 1999 by
Sutton Publishing Limited · Phoenix Mill
Thrupp · Stroud · Gloucestershire · GL5 2BU

This new paperback edition first published in 2003

British Library Cataloguing in Publication Data
A catalogue record for this book is available from the British Library

ISBN 0 7509 3208 2

Typeset in 10/11pt Photina.
Typesetting and origination by
Sutton Publishing Limited.
Printed and bound in Great Britain by
J.H. Haynes & Co. Ltd, Sparkford.

We are the pilgrims, Master,
We shall always go a little further,
It may be beyond that last blue mountain
        buried in snow,
Across the angry or glimmering sea.

These lines appear on the SAS's memorial
to its dead, in a churchyard in Hereford,
the SAS base.

'The Commando soldier . . . apte à tout!'

A French saying: 'apte à tout' = capable of everything.

CONTENTS

ABBREVIATIONS

AEDU	Admiralty Experimental Diving Unit
AIB	Allied Intelligence Bureau
AIF	Australian Imperial Force
CCS	Combined Chiefs of Staff
CGM	Conspicuous Gallantry Medal (all services)
CIGS	Chief of the Imperial General Staff
CO	Commanding Officer
CODC	Combined Operations Development Centre
COPP	Combined Operations Pilotage Party
COS	Chief of Staff
CQMS	Company Quartermaster-Sergeant
CSM	Company Sergeant-Major
DCM	Distinguished Conduct Medal (Army, other ranks)
DFC	Distinguished Flying Cross (RAF, officers)
DFM	Distinguished Flying Medal (RAF, other ranks)
DRG	Desert Reconnaissance Group
DSC	Distinguished Service Cross (Navy, officers)
DSM	Distinguished Service Medal (Navy, other ranks)
DSO	Distinguished Service Order (officers, all services)
Folboat	Raiders' canoe, probably from 'folding boat'
Gestapo	Geheime Staatspolizei
Goatley	Special operations canoe, named after manufacturer
G(RF`)	G(Raiding Forces)
HMS	His Majesty's Ship (now Her Majesty's Ship)
IO	Intelligence Officer
JCS	Joint Chiefs of Staff
LCA	Landing Craft Assault
LCI	Landing Craft Infantry
LCN	Landing Craft Navigation
LCP	Landing Craft Personnel
LCT	Landing Craft Tank
LRDG	Long Range Desert Group
LSI	Landing Ship Infantry
LST	Landing Ship Tank
MC	Military Cross (Army, officers)
MEHQ	Middle East Head Quarters
MGB	Motor Gun Boat
MiD	Mention in Despatches (all services)
ML	Motor Launch
MM	Military Medal (Army, other ranks, rarely to RAF and RN)
MTB	Motor Torpedo Boat
PM	Provost Marshal
POW	Prisoner of War
PRU	Photographic Reconnaissance Unit
PT Boat	Patrol Torpedo Boat
RA	Royal Artillery
RAAF	Royal Australian Air Force
RAF	Royal Air Force
RAN	Royal Australian Navy
RE	Royal Engineers
RM	Royal Marines
RMBPD	Royal Marine Boom Patrol Detachment
RN	Royal Navy
RNR	Royal Naval Reserve
RNVR	Royal Naval Volunteer Reserve
RSM	Regimental Sergeant-Major

ABBREVIATIONS

RTU	Returned to Unit	SRD	Special Reconnaissance Department
SAS	Special Air Service		
SB	'Sleeping Beauty', nickname for Motor Submersible Canoe	SSB	Special Service Brigade
		SSRF	Small Scale Raiding Force
		Sub	Submarine
SBS	Special Boat Section (Special Boat Squadron)	UDT	Underwater Demolition Team
SIG	Special Interrogation Unit	USMC	United States Marine Corps
SIS	Secret (or Special) Intelligence Service	USN	United States Navy
		VC	Victoria Cross
		X craft	Miniature submarine
SOE	Special Operations Executive		

ACKNOWLEDGEMENTS

For help and support of various kinds I must thank my late wife, Hazelle, my son Craig and my daughter Pirenne, and Anny De Decker. Chantal Persyn transferred a difficult typescript on to disk and I am grateful to her for her professionalism, as I am to Sarah Moore of Sutton Publishing for her skilful and sensitive editing. I owe much to the memory of Brigadier Peter Young, one of the greatest British Commandos, with whom I discussed Commando operations. Over many years I met and talked with hundreds of raiders who had served during the Second World War in SAS, SBS, LRDG and Commando companies. As an infantry soldier during that war I was not, strictly speaking, a raider within my own definition of that term, but from time to time I helped to train Commandos, notably those of Independent Companies of the Australian Imperial Force. Several Victoria Cross winners are mentioned in this book; full citations are given in my book *British VCs of World War Two*. Numerous people have helped me with illustrations; sources, where known, are shown with captions.

1

RAIDERS – 'REALLY KEEN MEN'

Raiding is one of mankind's oldest pursuits. Primitive man raided his enemies – and often his friendly neighbours – in order to steal women, slaves, land, food and better weapons. Since everybody engaged in this activity raiding was considered normal and among some groups of tribes it was the practice to take turns for a foray. Even the ancient Greeks, more sophisticated than most other peoples, lived by a well-established routine of reciprocal raiding. A leader's reputation was largely built on his skill as a raider and that same skill was a protection for his people, who might themselves be the victims of an attack. The tribe from over the hill could be reluctant to attack another if its chief had built up a reputation for sharp retaliation.

Down the centuries some peoples became famous – or more likely infamous – as raiders. The Vikings of the eighth to eleventh centuries were an example. They terrified the people of the east coast of Britain. Elsewhere, the Huns, the Tartars, the Vandals, Visigoths and Cossacks were among history's most notorious marauders.

Much nearer our own times, British sea captains, many of them pirates, daringly raided in various parts of the world, especially the French and Spanish coasts and the West Indies. Sir Francis Drake was little more than a glorified buccaneer who operated with royal approbation. Royal Navy captains of the late eighth and early nineteenth centuries liked nothing more than to raid French targets. But then the French themselves, as well as the Dutch and Spanish, were raiders of fearsome reputation.

Many British punitive expeditions, often involving seamen and soldiers, were raids in force. There were operations against native rulers who had reneged on some treaty or other or those who had dared to lay hands on a British envoy. Other attacks were made against slavers' hideouts and their slave-carrying ships. This was a principal activity of the Royal Navy in the northern parts of the Indian Ocean and Red Sea until well into the 1930s. Many a mission was vested with high moral principle – or so the organisers claimed – such as bringing Christianity to some benighted tribe or land. In the same way, Muslim attackers sought to spread the word of Islam.

In modern times the British armed forces have been the pre-eminent raiders, though not necessarily the most cost-effective operators, to use a twentieth-century phrase. Some raids were disasters, such as Dieppe in 1942, but many succeeded in their objective, including most of the SAS raids in North Africa, 1941–3. During the Second World War, the British carried out many more surprise operations than the armed forces of any other nation, without including bombing attacks made against civil targets. There seems to be something in the British psyche that craves the excitement of a raid against overwhelming odds, for this type of target is very dangerous. To my knowledge, only one major raid was made during the Second World War without the loss of a single raider's life. This exemplary venture was Operation Jaywick, with a British officer leading a party mostly comprising Australians. The target was Japanese-occupied Singapore in 1943.

In essence, Commandos were raiders and they carried out some spirited and successful operations, but as I describe in this book, in the later years of the war they were mainly used as conventional infantry, though in a spearhead role.

I define a raid as a surprise foray of limited duration made against a specific target such as a single enemy battleship, an airfield, a dock or a submarine pen. Many British exploits were carried out by two or three men, though in one major raid, Dieppe 1942, more than 5,000 Commandos took part. Italian underwater raiders operated in pairs and executed some spectacular attacks. Nazi German troops were successful in daring and imaginative raids – for instance, fewer than 100 men captured Fort Eben Emael in 1940.

I cannot give the Japanese kamikaze pilots the label of raiders. They deliberately sought suicide whereas raids of other belligerent nations were never suicidal – even though the tactics were sometimes so ill-planned and missions were so badly managed as to appear suicidal. Heavy casualties were inevitable. The Dieppe raid is again an example. Officers who were themselves not going on a mission which they had planned were profligate with men's lives. The planner who was also the leader in the field did his best to see to it that he and his men came through the venture. Some junior or relatively junior officers who planned and then led independent operations were among the finest military men of the war, fine, that is, in their achievements in action rather than in the rank they finally reached.

While the facts of history compel me to write more about the British raids than those of other nations, it has to be said that armies, navies and air forces of other countries had specialised raiders who performed well. The capture by glider-borne German troops of Fort Eben Emael, Belgium, in 1940, was the first raid of its type and one of the most dramatic and effective of all such exploits. Some of the United States raiding forces became famous, such as Merrill's Marauders, although their operations, like those of the British Chindits, should be more accurately called campaigns under independent command. The German Otto Skorzeny trained and led the best German raiders against targets of great significance or to achieve political victories, such as his rescue kidnapping of the Italian dictator, Benito Mussolini, in 1943. Russian commando squads struck again and again in attacks against German bases, small army posts, communications and transport.

Australia had commando units too: usually known as independent companies, they operated in the jungle-clad terrain to the north – Timor, Papua New Guinea, the Solomon Islands and Borneo. Two of the war's most successful raiding leaders were Australians operating for years behind Japanese lines in Borneo and the Philippines.

Some of the most effective raiders were the partisans or freedom fighters who operated in the German-occupied countries, especially in France and Yugoslavia, but I have restricted my accounts of operations and my analysis of certain ventures to raiders who came from regular military forces. Even so, some of these raiders operated as irregulars at times.

People, particularly those who played no active part in the war, are fascinated with raiding missions. In more recent times the fascination remains though in the wake of anti-war and pacifist movements some people regard these warriors as unprincipled killers. But what sort of man really made an efficient raider? Was there a blueprint by which men were assessed and then trained for this kind of warfare? What qualities did a raider need, what type of personality, what skills?

On 4 June 1940, immediately after the evacuation of the defeated British Army from Dunkirk, Winston Churchill, Prime Minister of Great Britain, made a defiant speech in the House of Commons during which he used a phrase which for him encapsulated the principal quality needed by men in the raiding units. Britain was woefully weak after the disastrous campaign in France and Belgium. While 330,000 men had been miraculously rescued from the French coast, many others had been captured, together with practically all the force's weapons, its armour, transport and artillery. Nevertheless, Churchill declared to the House, 'We will not be content with a defensive war.' This bold comment was typically Churchillian, aggressive and confident, even though Britain possessed nothing with which to be militarily aggressive.

In a letter to his right-hand man at that time, General Sir Hastings Ismay, Churchill was more specific. 'We should IMMEDIATELY [Churchill's emphasis] set to work to organise self-contained, thoroughly equipped raiding units.' Two days later, with the bit between his bulldog teeth, he wrote again to Ismay. 'Enterprises must be prepared with *specially trained troops of the hunter class who can develop a reign of terror* down the enemy coasts [my emphasis]. I look to the Joint Chiefs of Staff to propose measures for vigorous enterprise and ceaseless offensive against the whole German-occupied coastline, leaving a trail of German corpses behind them.' With such strong language Churchill was already setting out a pattern, though a simplified one, for Army Commandos. Implicit in his phraseology was his belief that the word raider was synonymous with hunter. Certainly the efficient raider needed the hunter's endurance and fortitude, his keen observation and an instinct for weighing the odds, the hunter's physical fitness and his 'survivability'.

The raider had to be willing to take calculated risks every moment when on a mission and he had to be ruthless. As a member of a small party among many enemies he could not afford to spare an enemy soldier because he was young, terrified and looked 'so innocent'. Anybody who might raise the alarm during a silent raid had to be quietly killed, whether by dagger, garotte or club. Often raiders could not afford the indulgence of taking prisoners. Attacking parties were few in number and each man had more than enough to do without having to guard prisoners, any one of whom might run for it or shout for help. In any case, prisoners being moved made a noise, when, by training and governed by the critical exigencies of the operation, raiders craved silence.

More than in any other branch of the forces, junior leaders had to be ready to take over a mission and direct it should their seniors be killed, wounded or captured, hence initiative above the ordinary was required of men who wanted to be members of raiding units. 'Wanted to be' is a significant phrase because in the British and Australian service such men were volunteers. This also applied to many raiders of other nationalities.

By volunteering, a man implicitly claimed élite status. In a way it also made him more expendable, since he was volunteering to run the risk of being killed. Unfairly, Commandos and men of the raiding outfits tended to look down on what they regarded as 'ordinary' infantry; they even tended to slight men of the Parachute Regiment, who were assault troops rather than raiders. Most raiders, Commando or otherwise, were required to be qualified parachutists but for them the parachute was merely a means to an end.

We can learn much about what was required of a raider by examining the demands of some of the officers who recruited them. Brigadier John Durnford-Slater, DSO and Bar, who claimed to be the very first Commando in British service, was responsible when he was a captain for forming the original Commando unit. 'I looked for the quiet, modest type of Englishman, who knew how to laugh and how to work,' he said. I am sure that by referring only to Englishman Durnford-Slater also meant Scottish, Welsh and Irish; he had men from all parts of the United Kingdom in his own ranks. He went on, 'I always avoided anyone who talked too much and I soon learned a lesson in this when a fine athletic-looking fellow who had taken part in many sports proved useless and boastful and had to be discharged. We [the Commandos] never enlisted anybody who looked like the tough-guy criminal type as I considered that this sort of man would be a coward in battle.' During the Second World War Commando and Raider leaders had the authority to return to their units any unsuitable officer or man. Some leaders made free use of this 'RTU' (Return to Unit) process during training. It was always humiliating but in most cases it was not a form of punishment. It was simply that in a raiding unit, especially among officers, there was no place for weak men or rotten apples.

Durnford-Slater also said that he was seeking 'men of character beyond the normal'. He considered that morale was the most important single factor making for success in war and that this spirit moved men willingly to strive and to endure. Durnford-Slater insisted that every soldier in the Commandos should be a potential leader, that he must be physically and mentally tough and must 'radiate cheerfulness, enthusiasm and confidence'.

In stressing all this, the Brigadier was at one with his great superior, Field Marshal Montgomery, who told me more than once that morale was the greatest factor in war. Montgomery considered that he was perhaps the greatest uniformed morale-builder of the war, and his leadership, especially in North Africa, seemed to support his claim.

Durnford-Slater aimed at making his Commandos the greatest unit of all time and by my assessment he came close to achieving his goal, at least during the Second World War. At his first meeting

5

with his hand-picked troop commanders he told them precisely what he expected of them, so it was not surprising that he got the best out of them. He was brutally frank. 'If you don't make the grade,' he warned them, 'it will be RTU.'

What David Stirling, founder of the Special Air Service (SAS) expected from his men, was just as revealing as Durnford-Slater's character 'sketch'. Stirling, one of the truly great soldiers thrown up by the war, almost certainly would not have been a success as a regimental officer and he was contemptuous of the lounge lizards who infested Cairo, but he knew what made a raider. 'Most of the work is night work,' he said, 'and all of it demands courage, fitness and determination in the highest degree. But also, and just as important, discipline, skill, intelligence and training. Many of these characteristics can be acquired. Training is designed to foster them. But this is possible only with really keen men. It is, therefore, no good men volunteering for this type of work for the novelty of it or for a change. We need the type of man who genuinely feels he has a special aptitude for work of this nature.'

Stirling's emphasis on work is interesting. He could have used 'fighting' or 'service' or 'duty', but for him raiding was a job of work, special work certainly but a labour to be mastered and enjoyed. Under Stirling and his lieutenants, all officers and men received the same training so that in the event of only one man of a patrol surviving he should be prepared to go ahead and finish the job himself. Unlike anywhere else in the forces, including the Commandos, he created a new relationship between officers and men, based on mutual respect and real friendship. That, with the ability to live with others, was essential when small groups were required to live, sometimes for months at a time, in the heart of enemy territory. As Captain D.I. Harrison, an SAS officer, said, 'Never before were the opportunities of individuality and team work so successfully welded together.'

Almost always, an officer's closest friend within the SAS was not another officer but an NCO or private soldier, his 'mucker' or mate. The officer was nominally the leader by virtue of his rank but he always sought the opinion and advice of his mucker. Survival depended on two heads being better than one and frequently a decision was reached jointly.

Raiders, regardless of the unit badge they wore, and whatever their rank, were highly self-disciplined, or they soon learned to be. All those officers who had clear ideas of the function of raiders insisted on this, though they relaxed the rules of formal services authority. An officer who wanted constant deference and blind obedience would almost certainly fail in a raiding outfit. John Durnford-Slater, David Stirling, George (Lord) Jellicoe, Paddy Mayne and all the other outstanding leaders were unashamedly élitist but for their men as much as for themselves. In wanting the 'finest fighting men in the world' they knew they had to 'pay' to get them and that payment was in the form of privilege. In the case of Commandos it was also financial; they received higher pay than other soldiers.

The finest fighting men in the world were never sorry for themselves. They almost exulted in times of stress, hardship and danger, because they knew they could take whatever the enemy or circumstance threw at them, whatever the lack of appreciation shown for their efforts. They had the innate mental toughness which, above all, made a raider. I cannot too strongly emphasise that raiding was not just a dangerous game. The attacks themselves were not merely tactical in their application, but strategic. Strategy is the intention, the desired result; tactics are the means by which the intention is realised. Simply put, the Allies' strategy in 1944–5 was to put so much pressure on German resources that Hitler and his generals would not have enough men, guns, armour and aircraft to deal with all the demands being made upon them – to the east (against Russia), to the south (the Americans, and the British together with Imperial troops) and to the west, the Normandy invasion (British, Americans, Canadians). The tactics were the battlefield means by which the strategic intention was brought about. The strategy behind the many operations carried out by the Special Boat Service in the Eastern Mediterranean in 1943–5 was to pin down large numbers of enemy formations so that the German High Command was less able to mount counter-offensives against British and American invasions. Handfuls of SAS and SBS raiders, sometimes in patrols of only two men, had an effect on the Axis war effort out of all proportion to their numbers. The Germans and Italians committed whole divisions – up to 20,000 strong – to fight the elusive British marauders.

Another major part of the overall strategy was to damage enemy morale and indeed German and Italian garrisons on the Mediterranean islands were terrified of the raiders. The tactics used to implement the strategy were not the business of High Command but of raider leaders on active service. Similarly, in the Western Desert of Egypt and Libya the SAS raiders were not merely indulging a taste for destruction and creating mayhem. They played an important strategic role. British air attacks on enemy airfields were costly to the RAF and sometimes the airmen, despite great courage and dash, failed to destroy a single enemy plane. Could the

SAS suggest a better strategy? Indeed it could. Using its specialised tactics of blowing up planes parked on airfields, petrol dumps, installations and spare parts dumps, the SAS achieved the results sought by the British Prime Minister and his generals.

Half a century after the end of the war people still retained the illusion that the SAS, SBS, LRDG and Commandos did what they did as a matter of military violence and vandalism, creating havoc for its own sake and as a means of giving themselves 'adventure'. The raiders did indeed bloody the enemy's nose but that was the strategy; the tactics were how they did it. Many people also regarded raiders with distaste and disgust as 'killers'. Killers they were – all fighting men were taught that it was their first duty to kill the enemy in combat. This was regrettable but inevitable. It was also often necessary for raiders to kill in order to save their own lives or to defend others, including the very people who spoke so critically of them. Some of their most vociferous critics appeared to have known nothing of the ruthless and murderous nature of the Nazi and Fascist enemies. It was they, after all, who massacred innocent civilians, who tortured their victims and butchered their prisoners as a matter of course. Raiders were not murderers but I have never met one who had any qualms of conscience about killing a uniformed enemy in the heat of battle or in setting a booby trap which would kill or wound. One raider for whom I have great respect, Captain John Lodwick, wrote of an occasion when an explosion turned enemy soldiers into 'strawberry jam'. This was distasteful, as Lodwick himself would have admitted had he lived into more mature years, but fighting men of his period thought in such terms. Many of his own comrades had been reduced to pulp in an explosion and a degree of callousness was a form of protection against grief.

There is one word I have not yet used in trying to establish the qualities that made a successful raider – audacity. The motto of the SAS – 'Who Dares Wins' – is a bold reference to the need for the audacious. Whatever the difference among nationalities, raiders of all countries shared this one quality. We might more crudely but more graphically say that they had guts.

If I appear to have concentrated on Army raiders in this book it is only because there were more of them than in the other services. The British developed naval raiders – above and below the waves – as well as parachute raiders and glider-borne attackers by air, to a higher degree than other races, though with some significant exceptions, such as the Italian underwater heroes and the German glider-borne intruders. Some raids were terrible disasters, others were outstandingly successful, such as the blowing up of the German battleship *Tirpitz* in a Norwegian fjord in 1943. In some operations heroism beyond belief was displayed for absolutely no advantage, such as that carried out at a house in Cyrenaica, North Africa, where General Rommel was supposed to be, in 1941.

All these events and many more had one thing in common – they were desperate ventures. Audacity in the name of some desperate endeavour was the quintessence of raiding.

2

INVENTING – AND TRAINING – THE COMMANDOS

While John Durnford-Slater may have been the first 'official' Commando, he did not 'invent' the British Commando idea. This is to the credit of Lieutenant-Colonel Dudley Clarke who at the time of Dunkirk, May–June 1940, was Military Assistant to the Chief of the Imperial General Staff, General Sir John Dill. A thinker worried about the future of his country, Clarke reflected on the wholesale defeat of the Allies in Belgium and France and asked himself this question: What can a nation do when its armies have been soundly defeated in the field but it does not accept this result as final? A student of military history, Clarke considered the Spanish guerrillas who had fought against Napoleon's occupation of Spain early in the nineteenth century. There was something to learn from their hit-and-run tactics against French camps and columns. Clarke had personally witnessed the Arab Revolt in Palestine in 1936 and he understood the near paralysis caused by the rebels' raids on British posts. Perhaps most pertinently, Clarke knew about the few thousand commandos, farmers in ordinary life, who had defied and humiliated a British Army of 250,000 during the South African war of 1899–1902. In the end the British and their Colonial allies had won but in terms of military skill and results in the field the elusive Boer commandos were the victors. There was surely a model here that Britain in 1940 could adopt, thought Clarke.

Late on the night of 4 June – the last day of 'the nine days of Dunkirk' – Clarke committed the fruits of his thinking to a single sheet of writing paper. The next day he submitted his cogent proposals to his chief, Dill. On 6 June, Dill took them to Churchill. Only two days later Dill informed Clarke that his scheme was approved. 'The PM believes that you have something here.' An unemotional man, Clarke was nevertheless openly elated. He was authorised to form Section MO9 at the War Office. The rapidity with which all this was achieved was brought about by Clarke's instant access to Dill and Dill's open door to Churchill. Clarke was fortunate that Churchill had the imagination to grasp the Commando idea and the power and energy to drive it through against the rigid minds of so many senior officers in the War Office and in the Army generally. Though Churchill had given Commandos his backing, the War Office and Whitehall blimps resisted the innovation, even though Britain was facing a crisis of survival with the Nazi enemy holding the upper hand everywhere. For years, proposals for units to be recruited and trained for raiding were rejected in high places and sometimes they were deliberately obstructed, as Durnford-Slater and Stirling were to find. Churchill, on the other hand, could see strategic and psychological gains. Along the occupied coastlines raids could raise the morale of the subjugated peoples and show them that they had not, after all, been abandoned to their fate. Massive help might be slow in coming but the raiders were its harbingers.

Clarke was ordered to mount an attack against a German target across the Channel 'at the earliest possible moment'. The instruction came from Dill but the language was that of the impetuous Churchill. In person, the Prime Minister told Clarke that certain conditions applied: in the face of imminent German invasion no unit could be spared from the defence of Britain and after the grievous losses of arms during the Dunkirk retreat the new Commandos would have to operate with the minimum number of weapons. On 18 June Churchill wrote a simple minute which clarified his thinking on the new Clarke proposals. 'What are the ideas of the C-in-C Home Forces about "Storm Troops" or "Leopards" drawn from existing units, ready to spring at the throats of any small landings or descents? These officers and men should be armed with the latest equipment, Tommy-guns, grenades, etc. and should be given great facilities in motorcycles and armoured cars.' This was Churchillian enthusiasm at its most vigorous and it demanded action.

Raising a raiding force was difficult because Army organisation was too inflexible to produce one. However, the decision to form a new style unit was taken. It would be called the 'Commando' and each sub-unit would be given a number. It is likely that Churchill accepted the idea of creating the 'Commando' from Clarke but the PM had also had many experiences of his own during the South African war, as a war correspondent, and he knew all about the frightening efficiency of the Boer commandos.

The Boer commandos had no problem with cumbersome and vulnerable lines of communication and supply – they did not exist. Each man travelled on a horse, or more likely, a pony. He was equipped with a rifle and bandolier, a bag of mealies and some *biltong*, strips of sundried lean meat. For extra rations the commandos slaughtered a sheep and they bought fresh vegetables and eggs from the native Africans. They were meticulous in paying for what they wanted; had they simply taken provisions they would have lost the sympathy of the people. The saddled-up weight of a farmer-warrior – the Boers hated to be called soldiers – was about 250 lb, while that of a British cavalryman was 400 lb. These dashing horsemen were crack shots and they knew how to use ground to their advantage.

David Stirling's SAS was a more lineal descendant of the Boers than were the British Commandos. The SAS – and the LRDG – had independent transport in the form of Jeeps and they carried with them all that they might need. The Boers each had an extra bag of ammunition; the SAS troops carried their rounds in boxes. They could stay in the field for up to three weeks, as the Boers did. In one way the Boers had an advantage when compared with the SAS and LRDG patrols: every farm and household was friendly, should they need shelter, food and fresh horses. In contrast, the British SAS units were out on their own. They certainly could not trust the isolated Arab communities in the deserts, even though a few proved themselves cautiously friendly.

For the new British Commandos, the first published notice outlining the conditions of service was terse. Commanding officers were told to send only their best officers and men to the new units, an order which was unlikely to be obeyed: all COs wanted to retain the best soldiers, not pass them over to somebody else. The officers and men must be strong and 'without medical blemish', which meant among other things that they could not have flat feet or short sight. They had to be young and able to drive motor vehicles. In 1940 the number of men able to drive was relatively small, other than in Army transport units, so this qualification narrowed the intake. Men who might become seasick were not wanted.

Recruits were found and Churchill had his way: the first Commando raid was indeed carried out as soon as possible. It took place on the night of 23/4 June 1940, only nineteen days after Clarke had presented his proposals. Major R.J.F. 'Ronnie' Tod led 120 men of the newly formed No. 11 Independent Company with the intention of making landings in the Boulogne – Le Touquet area. No great thought had been put into the selection of a target – what mattered to Churchill and everybody else was to make a strike against the enemy, anywhere.

Tod's force carried twenty Tommy-guns, fully half of all the Tommy-guns then available in Britain; this was some indication of

British impoverishment in the matter of modern arms. Using borrowed RAF rescue launches, the expedition left from Dover, Folkestone and Newhaven and came together in the Channel. Disaster loomed when pilots of RAF Spitfires on patrol thought the craft might be German patrol boats – nobody had thought to liaise with the RAF on the raid. The Spitfires zoomed down to attack and Tod managed to identify himself to the fliers just in time.

Tod's launch was leading the way into Boulogne harbour when a sweeping enemy searchlight picked out the raiders' boats. Tod at once turned south along the coast and the Commandos landed amid the sand dunes, where they brushed against a German patrol. The only casualty was Lieutenant-Colonel Clarke, who had gone along as an observer. A bullet nearly took off his ear. Ironically, the officer who had brought the Commandos into being was the first Commando to be wounded. South of Le Touquet another part of Tod's expedition found a large building protected by belts of barbed wire. They killed two sentries but were unable to cut through the wire so they threw some grenades into a window.

This first Commando raid was hardly one of dash and enterprise but at Dover the returning warriors were noisily saluted by every ship in harbour. Britain, though on its knees, had hit back. At Folkestone the reception for the returning warriors was less warm. Here it was thought that they were villainous German prisoners. When they explained that they were Commandos and that they had taken no prisoners, a port naval officer said, 'Commandos, you say. What's that?' And another officer mystified by the strange term, asked, 'Command what?'

The War Office made a brief public announcement that British troops had been in action in France less than a month after Dunkirk and public reaction was enthusiastic. On 14/15 July, Clarke's MO9 mounted another raid, this time against the enemy garrison of Guernsey. The result was a bloodless anticlimax: the landing craft were unsuitable and although Army and Navy did their job as best they could the mission was abortive. Churchill was so disappointed that he said, 'Let there be no more Guernseys!'

Despite Churchill's enthusiasm for Commando tactics and the publicity given to his favourable comments in War Office circles, the chiefs of the War Office regarded the Commandos as Churchill's private army and wanted them abolished, on the grounds that these units, which were 'unnecessary and irregular', were taking the best men. When the Commando officers sought a training area near Plymouth the Royal Navy, which regarded Plymouth as its fief, let it be known that creeping around at night slitting throats was 'so much nonsense'. Officials at the War Office tried dirty tricks in their campaign to get rid of Commando units. One of their ideas was to refuse to admit the existence of anything called 'Commando'. When they wrote to Durnford-Slater the envelope and the letter it contained were addressed to 'Lieut-Col J. Durnford-Slater, Officer Commanding a Company, Special Service Battalion'. Durnford-Slater met this tactic with even better dirty tricks. When a War Office letter was so important that it called for a reply he invariably signed his own letter 'Officer Commanding No. 3 Commando'. At other times he steamed open letters to see if they contained anything important. The routine correspondence he resealed and returned to sender with the rubber stamp UNDELIVERABLE – TRY NO. 3 COMMANDO. At the same time the colonel ordered his officers and orderly room staff that the term Special Service Battalion must not appear on any of No. 3 Commando's orders. In addition, in officers' messes which he visited he ran a campaign against the notion of a Special Service Battalion. He pointed out that the Nazis used SS as an abbreviation for the Schutz-Staffel, the units that were establishing a nasty reputation for brutality in occupied Europe. Britain could not afford to be tarred with the SS brush, he said; leave this stinking label to the Germans.

Within two months of beginning his campaign Durnford-Slater had his reward – the War Office surrendered. Their communications now arrived addressed to Lieutenant-Colonel J. Durnford-Slater, No. 3 Commando. Churchill would have been pleased; he had wanted his Commandos to show spirit and enterprise and they were doing just that. Churchill's influence on Commandos was felt in various ways. Durnford-Slater and his officers chose green for their Commando berets because of the PM's insistence that Commandos should be 'hunters' and the colour had an heraldic connection with hunting. Some senior officers considered that the steel helmet should still be worn in action but No. 3 Commando always wore the green beret. Durnford-Slater argued that the very light headwear gave everyone better physical ability and agility. He also said that he could think more clearly when wearing a beret than a steel helmet.[1] Once a man was returned to his mother unit he was no longer entitled to wear the green beret, which was whipped off his head by the RSM or Adjutant immediately the RTU 'sentence' was pronounced by the CO.

All Commando leaders insisted on cleanliness and tidiness, not only in quiet periods but in action as well. Durnford-Slater said that in the severest battle No. 3 Commando men always shaved and washed if a few drops of water were available. A wash and shave had nearly as great a stimulating effect as a good meal, he said.

Commanding officers, indeed all Commando officers, were always on the lookout for potentially good recruits and there was much competition for them. At one time Durnford-Slater heard that the Commando Training Depot at Achnacarry, Scotland, had taken in 600 police volunteers. He at once applied for 120 of them to be posted to No. 3 Commando. 'They were big, strong and intelligent and had all their police discipline and training behind them,' he commented. The policemen were the best single intake made by No. 3 Commando. Every man was a potential leader; many were later commissioned while others, with fine bearing and positive influence among the men, were promoted to sergeant. The quality of the recruits showed in No. 3 Commando's finest performance of the war near Termoli, Italy, in October 1943. At the head of the British advance against strong German positions, they were hammered by tanks, massively shelled by guns, attacked by infantry and were left exposed on their left flank. The Commandos held fast as other units crumbled around them. In the meantime Durnford-Slater called for every available mortar, tank and aeroplane to counter-attack the Germans. The Commando Brigade's action secured and held Termoli for him. General Montgomery, saved from having to fight for the line of the Biferno River, moved the whole Eighth Army forward. In addition, it gave Montgomery and his staff a valuable harbour next to the front line where men and supplies could be landed.

Obviously, the Commandos were not raiders in actions such as Termoli, but spearhead assault troops. Towards the end of 1943 Lord Mountbatten, as Chief of Combined Operations, had conceived the idea of enlarging the Royal Marine Commando role in the war. Until then there had been only one such unit, No. 40 Royal Marine Commando, an all-volunteer outfit along the lines of the Army Commando units. No. 40 had fought well at Dieppe and Termoli. Mountbatten wanted a mixture of volunteers and conscripts. Every Commando in the service opposed the idea; they were proud of their volunteer tradition and the officers declared that conscripted Marines could not be expected to maintain the high Commando standards. However, a general reorganisation of Commando units helped Mountbatten to have his way. The basic role of Commandos had been re-evaluated, so that they would no longer be raiders, pure and simple. They would land as an élite strike force and hold on until the Army with all its strength could join them.

In December 1942 the Commando Depot at Achnacarry, where all Commandos were trained, came under the command of Lieutenant-Colonel Charles Vaughan, who had earlier been second-in-command of No. 4 Commando. A First World War veteran, Vaughan had seen twenty-eight years' service in the Coldstream

Guards and the Buffs and there was little he did not know about the Army and soldiers. He was good at his job and nobody in higher authority ever considered moving him from it and he was still training Commandos at the end of the war. This was remarkable because the people responsible for posting officers to units so often took men who were capable and settled in a particular job away from it. This large, bluff man was said to have a warm heart and a sense of humour, but this was invisible to many of the Commandos who passed through his powerful hands. They gave him a number of nicknames, the most decent of which included Lord Fort William, the Wolf of Badenoch and the Rommel of the North. He liked Laird of Achnacarry best of all. More than 25,000 men trained at this centre – not only British but US Rangers, Frenchmen, Belgians, Danes, Dutchmen, Norwegians and Poles. Vaughan selected his staff with care. 'I do not want enthusiastic amateurs,' he often said to would-be instructors. 'I do want professionals. Are you professional enough to fire live ammunition close to my men, very close, without killing them?' He also told them, 'I don't care if you are unpopular but I care very much if you are not respected.'

When soldiers first arrived at the camp they were marched through an avenue of false gravestones – they were not told the headstones were phoney – the better to remind them that in a Commando's life one false step could mean death. From Day One live ammunition was used at Achnacarry. 'The enemy will not be firing blanks,' Vaughan told senior officers who felt that the risks at Achnacarry were too great. He expected his staff to produce diabolical experiences for the trainee Commandos. Thus, the 'Death Ride' was introduced – men crossed the Arkaig River by sliding down a rope. When a man fell into the river, which was often, he got out by himself. 'We do not have lifesavers here,' Vaughan said, 'and there will be none to fish you out of deep water on active service. If you do not actually want to drown, it's up to you so save yourself.' Of course, grenades were exploding close to the men as they made the 'Death Ride'. CQMS Bruce Frickleton arranged the 'Tarzan Torture', an obstacle that demanded strength and fitness, nerve, determination and raw courage. Frickleton, the chief physical training instructor at Achnacarry, reckoned that when a man survived 'Tarzan' he was ready to face the worst that Hitler, Mussolini and Tojo (the Japanese leader) could throw at him. By day, the Commandos ran up and down mountains, carrying weapons, ammunition and equipment and they were often soaked to the skin by the frequent rain of western Scotland. But they became fit and relatively few suffered from the ills of damp and cold. In the evening there was no respite from training and there was always time in the schedule for polishing the brass on their equipment, even though the moment they went on service everything shiny was taken from them. During the bloody Dieppe raid of 18/19 August 1942 a soldier is reported to have said to his officer, 'Jesus Christ, sir, this is nearly as bad as Achnacarry!'

Colonel Vaughan gave an address to the men of each course as they prepared to depart from Achnacarry. 'When you leave here you will go to civvy billets and get a special allowance, but don't imagine you get this for nothing. You will be on raids and operations. Some of you will be wounded, perhaps badly. Maybe you will lose an arm, a leg. I tell you now, you don't have to worry. You will be taken care of.' And then a pregnant pause. 'There will always be a job for you, up here at Achnacarry. We can employ limbless veterans.' Many are the stories about the realism of Achnacarry's training course. Some American Rangers, considering themselves already the best in the world, were apt to disobey instructions. They always paid a price. A Rangers officer was ordered to sit in the bottom of an assault boat during a river crossing exercise, but he regarded this as 'chicken' and perched himself on the side. Seconds later he was wounded in the buttocks by a bullet fired by an instructor. The American was then ordered to sit on his painful bum on the bottom of the boat until the exercise ended. Even then, with the wound dressed by a medic on the spot, he was expected to remain with his comrades until the party returned to quarters that night. To do him credit, he did not moan about the rough treatment but that evening he complained to a British officer who was a fellow-trainee. 'You were lucky,' he was told without sympathy, 'A German would have aimed at your skull, not your arse.'

The Ranger was very lucky. Of the 25,000 men who passed through Achnacarry's gate about fifty did not come out alive. They were drowned, accidentally shot or collapsed and expired from strain and exhaustion. Some died from pneumonia induced by the privations and extremes of temperature and exposure to which they were subjected. Some men who served as raiders with the SAS, SBS, LRDG and other élite units may have been through a course at Achnacarry but this would have been by chance and before they became raiders. Raider leaders trained their volunteers in the environment in which they would be campaigning, such as in the desert and in the waters of the Aegean Islands. After this they gained experience on the job. Some of David Stirling's early recruits, officers as well as men, had been Commando-trained and no doubt they were better soldiers for it. But they found that in the SAS – as in the other units – more was expected of them than in the Commandos. However, nobody shouted at them, as Commando NCOs had been prone to do. Stirling, his lieutenants and his successors never saw any necessity for parade ground bull.

The Commandos were reorganised into four self-contained brigades, one to remain in Italy, one to move to the Far East, and two for operations in north-west Europe. All were under a Commando Group HQ, whose first GOC was Major-General Robert Sturges, a Royal Marine. Durnford-Slater, promoted to brigadier, was deputy chief and in charge of the Army Commando units.

By the time all this happened, in 1944, Commandos no longer operated in the way they had done in their classic raiding years. Now they were something like super-infantry. One of the most interesting assessments of Commandos was made by Brigadier Peter Young, himself one of the most successful Commando officers of the war and a veteran of many raids and campaigns. His comments provide an insight that could only be offered by somebody who knew the Commandos from personal experience:

The Commandos of the 1940s may have been rather special in their way – they were all picked volunteers. But they were far from regarding themselves as anything out of the ordinary. Few were of gigantic stature and, until they received their specialist training, few were exceptionally skilful in the martial arts. The great majority had never even been under fire. They were just fed up with being told that the Germans were supermen and that they themselves were wet. And so they revolted against their age and went to war in a new spirit of dedicated ferocity. They rejected the lotus years.

The politicians of the League of Nations, of Disarmament and of Munich had lost their allegiance, if they ever had it. They revelled in the luxury of responding to uncompromising leadership in a cause that needed no explanation. They approached new tasks in a critical spirit. No tactic was sound just because the book said so. The men who rammed the lock gates at St Nazaire did a deed every bit as daring as the charge of the Light Brigade at Balaclava, but the old attitude of 'theirs not to reason why' was gone.

Not all senior officers would have been as laudatory of the Commandos as Young; some said that the Commandos tried to be larger than life. Having known Peter Young, I speculate that he would have responded, 'These boys had to be larger than life to do what was required of them.'

Throughout the war Commando soldiers were awarded 8 Victoria Crosses, 37 DSOs, 9 second DSOs, 162 Military Crosses, 13 second MCs, 32 Distinguished Conduct Medals and 218 Military Medals. This total of 479 is remarkable, even in an army where honours and awards were lavishly distributed. No. 3 Commando won at least 73 decorations out of the total of 479 – 8 DSOs, more than 30 MCs, 5 DCMs and more than 30 MMs.

1. I found the steel helmet so clumsy, uncomfortable and inhibiting that I contrived to 'lose' every helmet issued to me. The steel helmet jerked forward over the eyes when its wearer went to ground in a hurry, leaving him without vision for a vital second and causing him to raise his hand to steady the helmet and thus provide an enemy sniper with a target. I felt that if an enemy bullet were to kill me, then it would do so. I think that John Durnford-Slater had the same sense of fatalism. The steel helmet did have its uses. It made a good wash basin, water could be heated in it and rainwater could be caught in it. Inverted, it made a reasonably comfortable pillow, held by its chin strap it could be a close-quarter fighting weapon with the edge used in a slicing swipe. Small items of clothing, such as socks, could be washed in it. I saw soldiers urinate in their helmet but I could never understand why. I asked one man the reason he was employing his helmet in this way and he said, 'It's more modest than pissing in the open'. He was in the Army and he wanted modesty! However, I think he was serious.

3

THE SPECIAL AIR SERVICE

DAVID STIRLING DARED – HE WON

The story of how David Stirling managed to make a success of his concept of raiders work in the face of entrenched military bureaucracy is 'required reading' in order to comprehend his creation, the Special Air Service. The lengths to which Stirling was forced to go in order to translate some ideas scribbled on a piece of paper into one of the most successful military units in history illustrates the very qualities he demanded of the men who joined him. The motto given to the SAS – 'Who Dares Wins' – was the precept he followed in his determination to found his unit. For it was his, though he always gave credit to 'Jock' Lewes of the Welsh Guards, who was Stirling's first lieutenant and the SAS's earliest instructor.

David Stirling was one of the truly great soldiers thrown up by the war of 1939–45, if we assume 'great' to encompass enterprise, leadership, dynamism, endurance, courage and, above all, the desire to set an example. As a lieutenant-colonel at his most senior rank, Stirling was not great in the Montgomery or Eisenhower way, but actively in the field and as a leader of men in combat he had few peers during the war. In degree, though in a very different way, the German Otto Skorzeny was one of Stirling's few equals. In 1939, Stirling was a member of the regular Army, a 24-year-old second-lieutenant in the Scots Guards. With a height of 6 feet 6 inches he had a Guardsman's build too, but his mind was not cramped into the rigid mould common among his contemporaries. He was always contemptuous of the chairborne warriors who infested Cairo and who made life difficult for innovators such as he.

After the retreat from Dunkirk in June 1940, Stirling volunteered for service in one of the embryonic Commando units. Posted to No. 8 Commando, he came under the command of Captain Robert Laycock and with others, he trained in Scotland. No. 8 Commando became part of 'Layforce', named after Robert Laycock, now a colonel, and was sent to the Middle East. After a time it became evident to the impatient Stirling that he was not going to see the action he craved.

His boredom was alleviated to some extent when a friend, Lieutenant 'Jock' Lewes, another soldier with a restless nature, came across some parachutes in Cairo and 'appropriated' them; that is, he stole them. He asked Colonel Laycock for permission to 'experiment' with them, the experimenters being himself, Stirling and four other-rank Guardsmen.

Lacking an instructor, even without the help of somebody who had already parachuted, and in an aircraft unsuitable for the job, this adventurous group set out to learn a new military skill. They were young, enthusiastic and they gave little thought to risks. Their way of jumping from their borrowed Vickers Valentia, whose pilot they had bribed with the promise of a bottle of whisky, would have appalled any professional instructor. Normally, the static lines, which pull open the parachute of each parachutist as he jumps, are clipped to a cable running the length of the fuselage. Lewes, Stirling and party had no such refinement; they may not have known about it. They simply tied their static lines to the legs of metal seats in the aircraft. The principle may have been sound enough but as a method it was risky. When Stirling jumped his static line fouled, causing two panels of his parachute to be torn away. He descended rapidly, landed too heavily and damaged his spine so badly that his legs were paralysed. Rushed to hospital in Cairo, he spent two months under treatment.

In this enforced idleness he had time to follow the progress of the war in the Middle East. 'Progress' is not the best way to describe the British Army's misadventures. There was one failure after another, including General Wavell's Operation Battleaxe in June 1941, which was supposed to have relieved besieged Tobruk. The name Rommel kept cropping up as this brilliant and dynamic German general hammered his way from success to success. His very name and reputation badly affected British confidence. No British raiding force existed but it was badly needed in order to damage enemy morale and to help restore British fortunes. During June and July Stirling put his mind to the idea of raiding Rommel's long and obviously vulnerable supply lines, his airfields and supply dumps. He saw that an 'ordinary' Commando unit, no matter how brave and well trained its members were, was much too big for the type of independent warfare which he had in mind. Stirling's ideas were well formed. He proposed to raise a unit of 200 hand-picked volunteers who would be trained along lines which he had worked out. It would operate in five-man patrols. He put forward the revolutionary idea that these patrols could attack thirty enemy targets in a single night. Neither Commandos nor the very best infantry battalions were organised to do this.

To begin with, the greatest obstacle to the plan was the obstructive military bureaucracy in Cairo. Everything had to pass through what Stirling later described as 'layer upon layer of fossilised shit'.[1] Having committed his remarkable plan to paper, Stirling addressed his letter to the Commander-in-Chief Middle East, General Sir Claude Auchinleck. He proposed to hand it to the C-in-C personally but even major-generals had no direct right of access to this august person. Stirling took a taxi to the entrance of Middle East HQ and, with a friendly smile to the MP sentry, explained that he had left his pass behind. The sentry was polite but unmoved: nobody entered MEHQ without a pass. Even the crutches which this young officer was using made no difference. However, Stirling noticed a gap in the perimeter wire, just big enough for him to squeeze through. He left his crutches behind and set off at a fast limp for the building where he knew the C-in-C had his offices. The sentry blew his whistle, shouted angrily and gave chase. Inside the building Stirling chose an office at random as a bolt-hole. A red-faced major – Stirling's description of the officer – was startled by the intruder but apparently did not connect him with the shouting in the corridor. Stirling began to explain his mission in MEHQ but the major peremptorily ordered him out. To give himself more time for the hullabaloo to die down, Stirling patiently apologised before retreating to the corridor. He tried another door, pushed it open without knocking and found himself saluting no less a person than Lieutenant-General Sir Neil Ritchie, the Deputy Chief-of-Staff. He handed the general his written proposal which, in pencil, hardly looked impressive. Ritchie's first impulse may have been to call his aide and have the intruder expelled and that would have been the end of Stirling's military career. However, Ritchie accepted the paper, read it through and nodded to Stirling to sit down. He said that the proposal as outlined by Stirling could be just what was needed at that critical time and he promised to take it to Auchinleck. Stirling would have an answer within two days.

Having dared, Stirling had won. Ritchie was so impressed that on his own considerable authority he told Stirling to make a start with his planning. He would need help to do this and by telephone Ritchie called in another officer. It happened to be the red-faced major.[2] Auchinleck approved the Stirling proposals as readily as Ritchie had and a few days later Ritchie called Stirling in for discussions – this time he had a pass. He told Stirling that he had been promoted to captain and that he had authority to raise a force of six officers and sixty men. His operation must be planned with the Director of Military Operations at HQ. Stirling was not too keen about this 'interference', though he knew that he was not yet in a position to operate in isolation. Privately, he decided that he had to use the Army, not allow the Army to swamp him. Ritchie was a fast mover and he had already secured an area at Kabrit, near the Suez Canal, as Stirling's base camp. His outfit, Ritchie told Stirling, would be called L Detachment, Special Air Service Brigade. No such body existed, except in the fertile mind of Brigadier Dudley Clarke who was in charge of deception plans.

Stirling's first recruit was his friend Jock Lewes, who was already doing what Stirling was eager to do. Based in Tobruk, Lewes was making small-scale raids against enemy outposts. He was more than willing to join L Detachment. Stirling's second recruit was Blair 'Paddy' Mayne, another very big man, who had seen action in Syria with No. 11 Commando in June 1941. But there was a problem – Mayne was under close arrest for striking his CO in the mess. Close arrest meant that the defaulter had to be in the company of an armed officer of the same rank 24 hours a day and he was not permitted to leave barracks. With his usual soft and plausible effrontery, Stirling had Mayne released but there was a bigger hurdle to jump. Mayne was not impressed with Stirling, whom he had known as an idler in Layforce. In turn, Stirling knew of Mayne's personality defects. The two men, however, managed to come to terms and Mayne joined L Detachment. Before long he was second-in-command.

Mayne was continually in trouble with the authorities. His father had died and when he was called to Cairo for briefing about his new appointment he took the opportunity to ask for leave in order to return home for a time. When his request was turned down Mayne became sour and unruly. It was not difficult for him to find a target for his anger. He hated the many reporters in Cairo who wrote colourful and often wildly inaccurate stories about the SAS, the LRDG and other raiders without once getting to the scene of action themselves: David Stirling was often astonished to read some of the things which he was supposed to have done and said.

While on a drinking spree in Cairo, Mayne decided to beat up one of the most prominent journalists, the BBC reporter Richard Dimbleby. He blabbed about it, somebody tipped off Middle East Headquarters, which ordered the Provost Marshal to stop Mayne. The PM took six of his Military Police and they caught up with Mayne on the steps of Shepheard's Hotel, where Dimbleby was staying. The big, powerful and angry Mayne knocked out the PM, a major, and some of the Redcaps before the others could call for reinforcements. When the reinforcements came they handcuffed

Mayne and threw him into a military prison. MEHQ let Mayne stew overnight and then signalled the prison OC, 'Release this officer. He is more use as an officer than as an other rank.'[3]

With the nucleus of Mayne, Lewes and former Guardsmen, Commandos and sometimes other soldiers who were misfits in the conventional Army, Stirling vigorously set about training his unit. He was fortunate enough to locate several former members of the Commando troop he had commanded. On their first evening under Stirling's command, the raider recruits were told of their initial operation. It would take place that very night and the objective was to steal a camp for themselves – if they wanted a stretcher to sleep in and a tent for shelter. This was typical Stirling audacity. Frustrated and exasperated by the Army's tediously bureaucratic machine – and also by the wilful obstructiveness of the Adjutant General's Branch – from now on he took direct action to obtain what he needed. The New Zealand Division was absent from the Cairo area while fighting in the desert, leaving their large base camp deserted except for a caretaker squad. They did not caretake efficiently. Stirling's raiders lifted all the tents and equipment that they needed. Thoughtfully, they stole a piano for their own officers' mess. From this exemplary beginning Stirling's raiders knew that their chief expected them to be enterprising – in his terms, 'to dare'.

There was a significant difference between L Detachment (SAS) and the Commandos, from where most of the early recruits had come. L Detachment set a minimum but high standard of training which its members had to reach before being accepted. The Commando units, however, recruited their members and then somehow got them up to the necessary standard. Another difference was that the L Detachment men knew from the beginning that if they failed, for whatever reason, they would be classified RTU – Returned to Unit. Some Commando leaders, such as John Durnford-Slater, also adopted this policy but it was not general.

Commandos were taught to be 'tough' and encouraged to act tough, so much so that they exuded toughness for its own sake. They were not popular when on leave, especially when they were abroad. They wanted to demonstrate their macho image in bars and cafés and this resulted in objectionable behaviour. Many Commandos picked fights with civilians, foreigners, American troops and even with other British soldiers in a perverted display of their legendary toughness. This was by no means universal among Commandos and some commanding officers took a dim view of such behaviour. SAS, SBS and LRDG men, on the other hand, were expected to be correctly dressed and well behaved and in control of themselves when in public; officers strongly discouraged rowdy and aggressive behaviour. Paddy Mayne might not always have set a good example and when he drank too much he could be violent. However, in common with David Stirling, he advised his SAS men to keep their toughness for the enemy. Since many SAS men had come from Commando units they had to make a real effort to become the strong and silent type of soldier.

The SAS was unfortunate, perhaps, in its nomenclature. The use of the word 'air' seemed to indicate to outsiders, even high-ranking officers, that the SAS men were airborne troops. They were not and parachuting was only a minor part of their training. They were required to do the absolute minimum of jumps, just enough to give them the skill to drop safely. As Stirling and the other founding fathers of raiders knew, the SAS could be inserted into enemy territory by land, sea or air. Similarly, they were not trained to fight as a large formation; this was the job of infantry.

Some training was dirty, difficult and dangerous. Serge Vaculik, of Czech–French nationality trained in Scotland with 1 SAS and described his experiences:

We learned to fall into deep pits holding our weapons over our heads, and to lie face downwards on barbed wire while our comrades trampled forwards over us, and all along the course British machine-gunners, sharpshooters to a man, kept up a constant fire with live ammunition whipping up the ground a few feet ahead of us and behind us. If we were sometimes too exhausted to spring to our feet and dash on, the machine-gunners would treat us to a sustained burst of fire and that would lend us renewed strength to press on.

Stirling rarely sought authority for anything he wanted to do – he just did it. As all service innovators come to know, medium-rank superiors almost always say no to requests from juniors that are in any way 'out of the ordinary'. In this way, they hope to keep their noses clean, to avoid any mention of controversy in their file, and eventually to reach senior rank. Medium-rankers tend to play it safe. David Stirling knew that he was unlikely to be given authority for his unit to have its own badge, so he invented it. The badge was in the form of a dark blue cloth shield, emblazoned with a silver dagger with pale blue wings. Across the lower part of the blade, in a scroll, were the words 'Who Dares Wins'. It could have been the motto for every raiding outfit of the war. The parachute wings of the SAS, unlike those of the

Parachute Regiment, had a straight top edge and were woven in light and dark blue silk. In the centre and hard to discern was a small white parachute. Some of these wings were worn on the arm and some on the breast. The motto 'Who Dares Wins' has always been attributed to Stirling but the badge was designed by Sergeant Bob Tait, apparently in a unit competition. The wings were designed, so it was said, by Jock Lewes, but everything had to be approved by Stirling.

It was Stirling's own decision that all his men should be parachute-trained and in the Middle East he made the first drop himself, as has already been described. When he had qualified as a parachutist a man could put wings on his arm. When worn on the breast they were 'operational wings', which Stirling awarded for three successful missions behind enemy lines, though not necessarily by parachute. There came the day when Stirling considered that he had done really well so he issued the wings to himself. All this was highly irregular in an army which did everything by the book. On the steps of Shepheard's Hotel, Cairo, Stirling one evening encountered his Commander-in-Chief, General Auchinleck. The General returned Stirling's salute and then stopped in mid-stride. 'Good heavens, Stirling!' he said, 'What's that you have on your shirt?'

'Our operational wings, sir,' Stirling said.

'Well, well,' Auchinleck replied. 'And very nice, too; very nice, too.'

Stirling exploited this high-level compliment to prove to those officers who resented his success that the C-in-C himself had authorised the SAS wings.

Everything was in short supply in the Middle East in 1941–2, but not for the SAS. It seemed that nobody could supply Jeeps but Stirling obtained all that he wanted. Explosives could only be issued over the signature of a general but Stirling, without any official approval, somehow secured thousands of plastic and thermite bombs, used for creating instant raging fire amid enemy aircraft on the ground, and installations.

John Lodwick, one of the early raiders, said of David Stirling, 'He was the military Marks & Spencer. Like some vast organisation of chain stores his force grew . . . and grew . . . nourished by success, fortified by prestige and an intriguing air of mystery.' Stirling and SAS finally gained official recognition at high level in September 1942 when Major-General McCreery, Chief-of-Staff to General Sir Harold Alexander, analysed the future of special units in the Middle East. He reported to his chief:

The personality of the present commander of L Detachment SAS [Stirling] is such that he could be given command of the whole force, with appropriate rank. I make the following suggestions: L Detachment SAS Brigade, 1 SS Regiment and Special Boat Service should all be amalgamated under L Detachment SAS Brigade and commanded by Major D. Stirling with the rank of lieutenant-colonel.

At the same time Stirling's unit was given the status of a full regiment in the British Army, to be known as 1 SAS Regiment. This establishment came into effect on 28 September 1942. While Stirling was no doubt pleased, he was uncomfortably aware that this was a devious attempt to force him into more conventional military behaviour. He might have his regiment but now it came under the authority of the Director of Military Operations. This officer, a general, would control it through a specially formed HQ department – the G(Raiding Forces) (G(RF)). Nobody knew quite what this designation meant but the stipulation which accompanied it was painfully clear to Stirling: 'No other contact is permitted with the General Staff at GHQ, the naval C-in-C Mediterranean, including his staff, or HQ RAF ME, unless arranged by G(Raiding Forces).'

David Stirling had become accustomed to direct contact with these commanders-in-chief but now he had been put in a straitjacket. No longer could he approach senior officers at will or whimsy. To a large extent, he had worn out his welcome. In practice it made no difference to his style of working. When dealing with middle-rankers, such as lieutenant-colonels and colonels, Stirling managed to give the impression that he carried the authority of the high and mighty. 'The General wishes me to give you his compliments, sir', he would say to a less senior officer, 'and asks if you could help me with . . .' And here he would present his current 'wants' list.

After Stirling's capture, a black day for the British Army, the officers who had itched to get their fingers on his operations 'reorganised' the SAS, for lack of a better term. The Stirling-haters of MEHQ decided that the SAS was becoming too prominent and as a result various groups were split from it. Even the SAS's name was changed for a time, to Special Raiding Squadron (SRS).

Paddy Mayne was Stirling's natural successor to command SRS. Even in retrospect it is difficult to see who else could have done the job. Mayne had under him 250 officers and men and he divided them into three groups, commanded by Major Bill Fraser, Captain Harry Post and Captain F.H. Hornby. Like the Commandos, the SRS no longer operated in small independent patrols but as spearhead attackers in major actions. They lost little of their independent-minded spirit as inculcated by Stirling, but their further activities hardly qualify them as classic raiders as described in this book.

In May 1943 they were inspected by General Dempsey of 13 Corps, under whom they would do their campaigning. Dempsey spoke to the men in a camp cinema and instantly made a major blunder. Addressing the SRS men as if they were unblooded reinforcements, he said, 'I promise you your D-Day.' He meant that they would get their baptism of fire. There followed other well-meant comments but the seasoned warriors felt that they were being patronised. Being well disciplined, they listened to Dempsey but in grim and glowering silence. Paddy Mayne was standing with Dempsey and he now intervened and whispered in his chief's ear. According to witnesses, Dempsey was shaken by what Mayne told him and he made amends in a way that few generals would have done, for fear of losing face. 'Look here', he said, 'I've been giving you all this tripe about D-Day when you've had more D-Days than I will ever have.' He had cause to be angry with whatever staff officer had failed to brief him properly about the unit he was addressing. To be seen to apologise further, he stayed the night with the SRS and observed a night-firing exercise, which indicated to him at once that these men were not green.

Nevertheless, although Dempsey became very popular with the SRS and other raiders, the staff at HQ failed to understand their special skills and ethos and wanted to use them as shock troops. Stirling and his co-founders of the SAS had never envisaged such a role. Shock troops made spirited frontal attacks and suffered casualties. The purpose of raiders was to make stealthy approaches, to kill and destroy and then get out unscathed themselves. General Dempsey addressed the SAS again, in Italy. (I use the acronym SAS but the men were still the Special Service Brigade.) What Dempsey had to say provides interesting evidence of how the role of the unit had changed out of all recognition from that founded by Stirling, Jock Lewes and others. His comments also indicate the splendid service of these men, even if we concede that Dempsey, to a certain extent, was boosting their morale with compliments as generals have always done. He said:

It is just three months since we landed in Sicily and during that time you have carried out four successful operations. You were originally lent to me for the first operation – that of Capo Murro di Porco. That was a brilliant operation, brilliantly planned and brilliantly carried out. Your orders were to capture and destroy a coastal battery, but you did more. I left it entirely to you what you did after that and you went on to capture two more batteries and a very large number of prisoners. An excellent piece of work. No one would have foretold that things would have turned out as they have. You were to have returned to the Middle East after that operation but you then went on to take Augusta. You had no time for careful planning, still, you were successful.

Then came Bagnara and finally Termoli. [Here Dempsey refers to the Italian campaign.] The landing at Termoli completely upset the Germans' schedule and the balance of their forces by introducing a threat to the north of Rome. They were obliged to bring to the east coast the 16th Panzer Division which was in reserve in the Naples area. They had orders, which have since come into our hands, to recapture Termoli at all costs and drive the British into the sea. These orders, thanks to you, they were unable to carry out. It had another effect, though. It eased the pressure on the American Fifth Army, and, as you have probably heard, they are now advancing.

When I first saw you at Az-zib and told you that you were going to work with 13 Corps, I was very impressed by you and everything I saw. When I told you that you had a coastal battery to destroy I was convinced that it was the right sort of job for you. In all my military career, and in that time I have commanded many units, I have never met a unit in which I had so much confidence as yours. And I mean that!

Let me give you six reasons why I think you are successful as you are – six reasons which I think you will perhaps bear in mind when training newcomers to your ranks to your own high standards. Firstly, you take your training seriously. This is one thing that has always impressed me about you. Secondly, you are well disciplined. Unlike some who undertake this specialised and highly dangerous job, you maintain a standard of discipline and cleanliness which is good to see. Thirdly, you are physically fit and I think I know you well enough to know you will always keep that up. Fourthly, you are completely confident in your abilities – but not to the point of over-confidence. Fifthly, despite that confidence, you plan carefully. Lastly, you have the right spirit, which I hope you will pass on to those who may join you in the future.

Dempsey then explained that he had further points, which he always tried to bear in mind when handling the unit:

> These principles, if I may call them such, are: firstly, never to use you unless the job is worthwhile; that is, unless the effect to be gained is worth the risk taken in putting you in and there is always considerable risk when using troops such as yourselves. Secondly, never to put you in too far ahead of the Army. Always I must be able to reach you in twelve to twenty-four hours. If you are a small party, in twelve hours; if a large party at the most in twenty-four hours. Thirdly, I must be prepared to use the whole of my force, including artillery and tanks if need be, to reach you within that time. One reason is that you always seem to stir up trouble wherever you go. [In this, at least, the SAS were still David Stirling's boys.] Fourthly, I always try to give you as much time as possible for careful planning. On the other hand, I bear in mind that I must not hesitate to use you quickly if the opportunity suddenly arises. Such a case was Augusta and you succeeded as only a well-trained unit would succeed. Finally, once you have carried out your job I must get you out as quickly as possible to enable you to refit and to reorganise.'

It was not surprising, after this catalogue of the SAS's merits that Dempsey was forever the unit's favourite general. Paddy Mayne, in command of these military paragons, would have been pleased by Dempsey's compliments and he realised that Dempsey had now fully redeemed himself after his unfortunate gaffe months earlier, when he had patronised the unit.

Like his men Mayne was an enthusiastic but discriminating looter. Apparently, his special interest was cameras, while other SAS/SRS men fancied pistols and revolvers, binoculars – the Germans made the best binoculars in the world – watches and jewellery. Most raiders had a fine taste for booty and would never overload themselves with bulky or weighty items.

In his first action in Italy as CO of the SRS Mayne won the second of his four DSOs. The citation for this award says much about his leadership:

> . . . The operation was to capture and hold the town of Augusta. The landing was carried out in daylight – a most hazardous combined operation. By the audacity displayed the Italians were driven from their positions and masses of valuable stores and

equipment were saved from enemy demolition. In these operations it was Major Mayne's courage, determination and superb leadership which proved the key to success. He personally led his men from landing craft in the face of heavy machine-gun fire. By this action he succeeded in forcing his way to ground where it was possible to form up and sum up the enemy's defences.

In most of Mayne's citations and mentions in despatches it was noted that he totally ignored his personal safety. The reference to an attack under fire from landing craft clearly indicates the SAS's change of role. It was completely alien to the original ethos and function of Stirling's raiders.

Mayne commanded the SAS in Europe, 1944–5, but again they were not raiders in the way Stirling would have defined the term. Nevertheless, they made forays against the Germans, often in concert with the French Resistance. Throughout this period the SAS was immensely successful, yet at no time did the brigade number more than 2,000 effectives. Of these 330 were killed, wounded or reported missing. Many of them were murdered by the Nazis, following Hitler's directive that Commandos were to be shot.

In 1944/5 the brigade claimed 7,733 enemy killed and 4,784 made prisoner and they acknowledged that Resistance groups led by SAS officers accounted for a proportion of these enemy casualties. The destruction was immense: 29 locomotives, 89 wagons and 7 trains in their entirety; 33 derailments with lines cut at 164 places; 740 vehicles destroyed or captured; 400 bombing targets reported to the RAF.

The raiders were audacious to the last, even when they returned to Britain. A tank landing ship ferried SAS troops, still with their Jeeps, to the port of Tilbury. From the ship, the observant SAS men noticed that a large contingent of Customs officers was waiting for their arrival and elementary speculation convinced them that these hated officials were out to impound the large quantities of cigarettes, liquor and precious booty so painstakingly and lovingly collected during active service. The SAS was not about to surrender its treasure to Customs or to anybody else.

Somebody, almost certainly Lieutenant-Colonel Paddy Mayne, had an idea – and if it was not his idea it should have been. The SAS men sat in their Jeeps behind the LST's great bow doors, the drivers steadily revving up their engines. As the steel doors descended and even before they clanked down into place, the roaring Jeeps took off in rows. The Customs men, grouping for their own raid, had to jump for their lives as the Jeeps, some of them airborne, landed with squealing tyres and dispersed into the Essex countryside. Customs did not lay its hands on so much as a single bottle of contraband liquor. It was an operation worthy of David Stirling himself.

1. He was speaking to Anthony Kemp, author of *The SAS at War 1941–1945*, and researcher *par excellence* into SAS history.
2. In 1957 Stirling told the American journalist Virginia Cowles that the major had been one of his instructors at the Guards Depot in Britain and that he had slept through his boring lectures. The major had not forgotten this discourteous tall young Scots Guards officer.
3. Anthony Kemp quotes this story in his *The SAS at War 1941–1945*. Elements of it came to him from Reg Seekings, one of David Stirling's most devoted followers.

4

BY THEIR DEEDS SHALL YOU KNOW THEM

SOME SAS RAIDS AND FEATS OF ENDURANCE

A full description of all SAS raids would fill several volumes but nobody could write it because the details of many raids do not exist. In a lot of cases, they never did exist because the raiders were too busy or too tired to write reports or they were simply not interested in putting pencil, pen or typewriter to paper. David Stirling was a poor records-keeper, as was Paddy Mayne. At the end of the war nobody in the raiding units gave any thought to the historians who might one day want to write about the operations. The officers charged with sending paperwork to some HQ or other just crated the files, addressed them and gave them no further thought. Archive keepers somewhere received these masses of paper and stored them.

Much SAS material finished up with Army Air Corps Records Office, Edinburgh. At some time down the years the 'weeders' went to work and shredded or burnt tons of it. Official 'weeders' are among the most evil people on earth; they deprive posterity of much intriguing history. On what basis they do this 'refining' – that is the official description for this vandalism – is impossible to say.

Even the Public Record Office (PRO) has little material coded SAS but indirectly, it is possible to gain an oblique focus on the activity of the raiding units through the study of documents which were kept as unit, formation and HQ war diaries. The most useful sources are the accounts of missions written after the war by officers and men who served in the units, newspaper articles of the time and interviews recorded by correspondents active at that period in the Middle East, southern, western and north-west Europe, and North Africa. Some authors went zealously to work, none more enthusiastically and thoroughly than Anthony Kemp, author of *The SAS at War*.

I have selected some of the more interesting SAS and SBS raids to illustrate the daring, bravado and sheer courage of the raiders. Sometimes their confidence overrode their better judgement.

PADDY MAYNE AND THE TAMET RAID, NOVEMBER 1941

Paddy Mayne, an Ulsterman, was one of the most formidable British soldiers of the war. Nothing daunted him and he had the hunter-killer streak that Churchill had identified as perhaps the most fundamental quality of the raider. Of Mayne's many raids in North Africa, perhaps his exploit at Tamet (or Tamit) is the one which best illustrates his style.

Tamet, a coastal town in Libya, 30 miles west of Sirte, was known to have a new enemy airfield. Operating from the oasis of Jalo, David Stirling was seeking worthwhile targets and he planned to take on one objective while Mayne, with nine men, targeted the enemy aircraft, mostly Italian, at Tamet.

Mayne had already pointed out to his SAS comrades that while it was fine to blow up aircraft on the ground, this still left trained aircrew who could fly replacement planes. If crews could be killed, he said, the SAS would be striking a double blow against the Nazi–Fascist enemy.

At night, dropped by an LRDG patrol at their start-point, Mayne and his men set off in single file and met no obstacles during their approach. At first they saw nothing in the black cold of the North African winter but then they came upon a row of buildings which lined one end of the airfield. Seeing a strip of light at the bottom of a door in a prefabricated hut, Mayne crept up to it and heard men talking and laughing inside. He took the hut to be an officers' mess.

This was a choice target. As Mayne flung open the door his eyes took in a party-like atmosphere and, he thought, about thirty people were present. His interruption had been noisy and everybody became silent and looked towards the doorway. They saw a big man wearing British battledress and carrying a Tommy-gun at the ready. Mayne sprayed the room with his gun, saw men fall and shot out the lights with his final rounds.

The shocked survivors in the hut began to fire wildly through the windows. Quickly, Mayne ordered four of his men to continue this fight – 'in your own way'. He led the other five men on to the airfield where they deftly placed time bombs. They used time-pencil fuses, which worked on the principle of acid eating through wire, which, when severed, set off the explosion. These fuses generally worked but they were not always accurate, as the timing was crudely determined by the thickness of the wire.

Exasperated when he found he was one bomb short, Mayne climbed into the cockpit of the remaining plane. Some of his men, accustomed to their leader's sometimes eccentric behaviour, wondered if he was about to take off. Then they saw him haul himself out of the cockpit with the entire instrument in his powerful hands. By sheer force he had wrenched it from its fastenings.

From the lack of firing in the officers' mess, Mayne knew that his men had dealt with the enemy and were making their way to the rendezvous with the LRDG, as ordered. With his five men, Mayne now headed in the same direction. At that moment the first plane blew up and other explosions followed. The SAS men had to discipline themselves to press on rather than stand still and enjoy the satisfying spectacle. Even so, they were spotted against the light from the blazing planes and somebody shouted 'Avanti!' at them. Mayne responded to this challenge by crying out 'Freund!' but the grenade that he flung with his reply showed that he was anything but friendly.

The fireworks were being observed by the LRDG patrol a few miles away. 'Paddy's lit another bonfire,' one of them said. They flashed lights, as planned, to guide the raiders to safety. The Italians were also flashing lights and they were confusing to the SAS men. Mayne used his back-up signal – blasts on his whistle – which was answered by the LRDG men. Mayne had destroyed twenty-four aircraft and killed an undetermined number of enemy without a casualty among his own raiders. On another occasion, before destroying planes on an airfield, Mayne and his men killed the enemy aircrew while they slept. If it seems 'unsporting' to kill the enemy while they are asleep, consider this: what was Mayne to do with them if he took them prisoner? He had no spare men to guard captives and certainly no transport for them. Also, it was preferable not to leave wide-awake enemy who could describe the raiders.

A distinguished professor without war experience once said to me, apropos Mayne's raid in which he killed sleeping men, 'At least he could have woken up these men.'

'And then what?' I asked. 'Was he then to kill them or tell them that they had five seconds in which to go for their firearms in a shoot-out with the British? The raiders could have lost their own lives or Mayne may have wounded to care for instead of getting on with his mission. In fact, the mission would be at an end.'

'I see all that,' the professor said, 'but in cold blood. . . .'

'There was no such thing,' I explained. 'The blood of the SAS men was up. They were behind enemy lines, without possibility of rescue. They were at war, not playing games.'

Paddy Mayne destroyed more aeroplanes than any other soldier in the world. Only fighter pilots continually strafing enemy airfields could have demolished more. It was almost as if he had a phobia against aircraft and felt a psychological compulsion to wreck them. But Mayne was a rational man and his hatred was for the Nazis and Fascists. With every plane he destroyed he was saving British and Allied lives.

It needs to be said that he and Stirling were not the only SAS 'stars' in aircraft destruction. Soon after Mayne's Tamet raid, Lieutenant Bill Fraser, with Sergeants Tait and DuVuvier and two privates, destroyed thirty-seven aircraft on the airfield of Agedabia. Fraser and his men were still planting bombs while their first charges were going off and they disregarded the wild shooting of the frantic Italians. Again, their LRDG standby team were excited witnesses of the noisy and vividly lit exploit. Fraser, as a major, was still raiding German targets in Germany itself in 1945, during the final months on the war.

STIRLING'S BENINA RAID, LIBYA, 8 JUNE 1942

As one of a series of raids mounted by his group, Stirling himself with his two inseparable soldiers, Reg Seekings and Johnny Cooper, drove to their target, Benina airfield, near Benghazi. The town was being bombed by the RAF at the time so the sentries were distracted. What happened on this raid and immediately afterwards showed not only the benefits of audacity but the risks which accompanied an urge to go too far and do too much. This was in Stirling's nature and in June 1942 it nearly finished him.

The airfield was in an uproar after the RAF's raid as Stirling and his men found a place to hide among the buildings. Here he gave Seekings and Cooper a talk about deerstalking in the Scottish Highlands, one of his keen pre-war pursuits. By now Seekings and Cooper were accustomed to the extraordinary coolness of their officer so they listened attentively while he drew parallels between stalking game and outwitting Germans.

When the time was ripe and the garrison was relaxing after the period of high tension, Stirling and his men got down to work. Using time pencils, they placed bombs on two aircraft in the open and on several more planes and spare engines in hangars. On the way out, Stirling noticed the guardroom. He kicked open the door and tossed in a grenade with the quiet invitation, 'Here, catch!' The grenade exploded and the trio ran for safety as the aircraft began to blow up.

From a hiding place atop an escarpment, they watched in satisfaction as their sixty bombs continued to cause havoc at Benina.

Meanwhile Paddy Mayne and his party had been busy nearby. For once, Mayne was unable to blow up aircraft but by a ruse he managed to induce two parties of guards to fire furiously at each other. Making the pre-arranged rendezvous, Mayne and Stirling with their patrols drove down from the escarpment to view the previous night's destruction and at the same time shoot up anything 'worthwhile' on the Benghazi road. They were quite carefree as they approached a roadblock, which was manned not by the usually gullible Italians but by half a platoon of alert Germans. With Stirling and Mayne was Karl Kahane, a Palestinian Jew with experience in the German Army; it was now up to him to talk Stirling and his raiders through this crisis.

Kahane told a German sergeant that they were coming in from a long patrol, during which they had captured the Chevrolet truck in which they were travelling. The password? Kahane knew only the password for the previous month but again he explained that his group had been in the presence of the enemy for six weeks and out of touch with their own unit. The NCO seemed to be satisfied but he heard Mayne, on the other side of the vehicle door, cock his automatic. Other such ominous sounds told the NCO that he could die then and there or wave the Chev through. He chose the survival option.

By the time the British reached the next roadblock the enemy had set a trap. Italians were at this post and they excitedly waved their rifles and screamed orders to stop. Mayne, who was driving, shot through the Italians while Stirling opened up on them with a Vickers machine-gun mounted on the back. The SAS men were looking for targets but the only one available was a petrol-filling station where numerous trucks were parked. Rapidly, the raiders laced them with time-bombs and the party went hell for leather into rough country. They evaded enemy patrols but their truck was badly mangled on the stony tracks. Everybody was exhausted by dawn but they were galvanised into action by a shout from one of the men, 'Everybody out! Bomb!' The jolting had activated one of the bombs and it would explode in 20 seconds. Nobody was hurt in the blast but the Chev was wrecked.

Stirling, Mayne and party returned to base without casualties but they had pushed their luck and they were fortunate that enemy spotter planes, though directly above them, did not see them concealed in the scrub.

JELLICOE'S RAID IN CRETE, JUNE 1942

The SAS not only plagued the Germans and Italians with raids in North Africa but in occupied Crete as well. It was here that Captain George (Lord) Jellicoe, in the company of Commandant Georges Bergé of the Free French Army, carried out an audacious raid. The patrol was after sixty-six enemy aircraft, mostly Junkers 88 bombers, on a closely guarded airfield. Each Junkers wrecked meant fewer bombs dropped on British troops, airfields and ships and, even more importantly at that time, on battered Malta.

Jellicoe, aged only twenty-three, had inherited the title from his famous father, an admiral during the First World War. Formerly a member of the SBS, he had transferred to the SAS only a month before his Crete exploit and was not considered fully trained. He was junior to Bergé but on this venture they were, in effect, in joint command.

Jellicoe had proved himself to be brave but when there was any mention of his courage, he said offhandedly, 'Put it down to benzedrine.' He was not the only raider to take this drug, which was helpful as it kept men awake when it would have been dangerous for them to sleep.

The Bergé–Jellicoe team consisted of a Greek guide, Costi, and three French soldiers, Leostic, Sibert and Mouhot. Leostic claimed to be eighteen but in fact he was only fourteen. The submarine *Triton* took the party on a four-day cruise from Alexandria and landed them on a beach near Heraklion. Jellicoe and Bergé needed their benzedrine because the group marched all night through the mountains. They hid in a cave during the day and walked all the next night as well, reaching their target airfield about three in the morning.

Even at that hour the airbase seemed to be active and as the SAS men moved cautiously forward a sentry shouted a challenge and fired a shot as the raiders scattered. No general alarm was raised and they walked on for some distance looking for a way into the base. Again they encountered a wide awake sentry who, though on the alert, did not hear them. With only an hour to daylight, Bergé and Jellicoe wisely decided to make their attack the next night.

From a good hideout on a cliff they studied the airfield for most of the day and felt more confident for this reconnaissance by eye. They began their approach at 10 p.m. and were close to the perimeter wire when a patrol came marching in their direction. The five raiders dropped into bushes, hiding their Tommy-guns under them. When the patrol passed their tension eased but the last member of the patrol ran back, shining his torch. In turn, the team picked out Bergé, Jellicoe and the other three. The soldier asked what they were doing. They could have opened fire but that would have ruined their raid. One of the French soldiers, Private Mouhot, a Breton, feigned what Jellicoe later described as 'a ghastly lingering, drunken snore' and followed it with a belch. Such antisocial noises were apparently normal on Crete and the German – and some of his comrades who had joined him – were satisfied that these strangers sprawled in the bushes were stupid Cretans sleeping off an alcoholic stupor. Making some disparaging remarks, they moved off. Once again a simple bluff had paid off.

Giving the patrol a few minutes, the SAS men cut their way through the perimeter wire and hid in a bomb dump. This concealment was just in time for the Germans came hurrying back and discovered the hole in the fence. The raiders heard alarmed voices and Jellicoe feared that they might have to shoot their way out.

At that critical moment the SAS men got lucky. A flight of eight Junkers 88s roared in low to land and their arrival attracted the attention of the entire base. Jellicoe saw that a lone British Blenheim bomber was closely following the Junkers. Its pilot, clearly a courageous opportunist, made the correct recognition signals, which deceived the enemy into thinking that he was one of their own bringing in another German plane. As the Junkers landed the Blenheim dropped its bombs from low level. The crew's aim was so poor that they missed the airfield altogether but the explosions caused great commotion on the base and in the confusion the SAS men left their hiding place and glided into the darkness. They placed bombs in the enemy dump close by, then in sixteen Junkers and some way off they dealt with six more. It was time to leave but on the way out they tagged behind an enemy working party and 'laid eggs' – as some raiders used to say – on yet another plane, on four trucks and a petrol dump.

To give themselves time to get clear, they had used two-hour fuses, so they were well on the way back to their hideout when the explosions began. Clouds reflected the flames of the burning petrol and planes. The SAS men were now high enough to see some of their destruction in the light of the breaking dawn. Jellicoe, usually a calm man, later admitted to hugging Bergé with delight.

At dawn they reached the rendezvous with Costi, their guide. Now the whole region was in uproar with spotter planes, motor patrols and troops searching for the attackers. Relatively safe more than 12 miles from the scene, the SAS men hid all day and that night Costi led them to within a few miles of their escape beach.

They would hide for another day and keep rendezvous with *Triton* the following night.

A crisis occurred. Some Cretans, probably out game hunting, came upon the SAS raiders, took them for Germans and bolted in panic. About an hour later another man, acting in a friendly way, turned up and offered his help. Costi knew him well so he seemed to be safe and he went off to obtain food for the hungry men.

Soon after his departure, Jellicoe and Costi separated from Bergé and the Frenchmen to go to a village where they believed that they would find a certain Cretan, whom Jellicoe knew to be in the pay of a British spy. Jellicoe hoped to contact the spy who would signal *Triton* when to be ready to pick up passengers. Jellicoe and Bergé arranged to meet on the beach. But the friendly Cretan who had offered his help in the morning failed to return and the uneasy Bergé ordered his men to move. It was soon all too obvious that the Cretan had betrayed them. Bergé saw strong patrols of German soldiers approaching from east and west when he changed direction south he spotted a third patrol in that direction. He was being squeezed into a trap.

It was mid-summer and Bergé reckoned that he had to survive for an hour more of daylight before darkness would give him a chance to escape. Until then, he and his men had to fight, even though their Tommy-guns were useful only at close range. The Germans fired machine-guns and lobbed mortar bombs. When one enemy soldier ran towards the SAS men they shot him dead. For some minutes nothing happened apart from the odd mortar bomb explosion. Bergé reckoned that the Germans preferred to take him alive for interrogation.

Then the nerve of young Leostic broke and jumping up he exclaimed breathlessly, 'Mon commandant, je ne peux plus vous obéir!' [I cannot obey you any longer]. Running out to get a better position to fire at the Germans, he was hit by bullets and died soon after. Bergé and his remaining men ran out of ammunition and Siberty, wounded, was taken prisoner. Bergé and Mouhot changed hiding places but were found and captured. A rubber boat from *Triton* picked up Jellicoe and Costi on time at 10 p.m. Jellicoe later transferred back to the SBS, which he came to command. Bergé, though roughly treated, survived the war.

Jellicoe was praised for his performance but his superiors realised that a target of sixty-six enemy aircraft required a larger patrol than five raiders and a guide.

STIRLING RUNS AMOK AT BAGOUSH AND SIDI HANEISH, 26/7 JULY 1942

At Bagoush airfield 40 miles behind enemy lines, Stirling and his team evolved a new, daring and violent form of raiding attack. Mayne had bombed Bagoush the previous night but was disappointed with the effect, apparently the result of some bombs failing to explode.

Stirling decided to make a naval-style attack. Setting the pace in his Blitz Buggy, he led several Jeeps, all equipped with twin Vickers guns, in a blazing assault, like so many motor torpedo-boats on the loose amid an enemy fleet. The SAS ran amok on the airfield, shooting up and setting fire to numerous aircraft, installations and barracks. The defenders were so demoralised by the high-speed attack that they were capable only of wild firing and not one of the SAS men was wounded.

The success of the Bagoush Jeep raid fired Stirling's imagination. He wanted demolition of enemy aircraft on a grand scale and Sidi Haneish airfield came to his attention as a probable target. He rehearsed his Jeep crews in the tactics of the raid and on 26 July the final dress rehearsal took place many miles behind enemy lines, a classic Stirling touch of audacity. The Jeeps, in two lines of seven, would drive straight on to the airfield where the targets would be at the discretion of the column commanders, Paddy Mayne and George Jellicoe. Both columns would fire outwards so as not to endanger comrades. Stirling would lead up front while three other Jeeps, as spares, were ready to take over from any that might be wrecked. They also carried more ammunition and some medics.

On the evening of 26 July the Jeep columns set off in formation to drive something like 40 miles to the target. As they approached the airfield its floodlights were switched on and Stirling, suspecting that he had been spotted, abruptly halted the advance. At that moment a German bomber came in to land; the lights had been for the pilot. Stirling took advantage of them and led his Jeeps towards the runway, with all twenty-eight Vickers guns blazing.

The terrified Germans and Italians switched off the lights but within a minute the flames of burning aircraft and fuel dumps lit up the entire area and provided all the light that the raiders needed. The SAS attackers came under fire but they moved so fast amid the glare that the enemy gunners mostly missed their targets; Stirling's Blitz Buggy was one of the few hit but another Jeep picked up him and his crew.

That night the SAS destroyed forty aircraft, many of them vital Junkers troop transporters. The Luftwaffe and the enemy defences suffered a serious reversal. The raid might have been even more destructive had not some of the SAS gunners been too enthusiastic in their firing and run out of ammunition during the raid. Stirling lost one man killed and a handful of men lightly wounded.

Dispersing as ordered, all crews returned safely to their camp. As never before, Stirling's tactics were vindicated by this raid. The great offensive under General Auchinleck had faded away inconclusively and only the SAS had been unreservedly successful. It is not surprising that the unenterprising and jealous senior officers managed to see to it that Stirling was not decorated. His exploits merited the DSO.

SAS RAIDS TO SUPPORT MONTGOMERY, TUNISIA 1943

Stirling understood the strategic picture in North Africa better than almost anyone and because of his ambitions to tackle the situation in 1942 he was planning, for the future of the entire SAS Regiment, as it now was. It had a strength of 83 officers and 570 other ranks and could no longer be considered a 'marauding band of desperados', as one staff officer had earlier called Stirling's men. And the SAS was growing even larger with a second regiment being raised in Algeria by Stirling's brother, William.

Stirling thought big as did his chief, General Montgomery. The general had been offhand when he met Stirling – very likely because he was short in stature and Stirling towered over him – but Stirling had his boss's best campaign interests at heart. Montgomery proposed to attack Tripoli and push on into Tunisia in an operation that was to begin on 15 January 1943. Eighth Army HQ, now more cooperative than before, accepted Stirling's proposal for SAS actions to support the main assault.

In brief, Lieutenant Harry Post, leading a Jeep patrol, was to play hell with enemy lines of communication west of Tripoli, that is, behind the German–Italian lines. Lieutenant Jordan, with a strong Jeep force, would raid convoys using the Gabes Gap. At Montgomery's direct request, a third patrol was to reconnoitre the enemy's Mareth Line so that Montgomery would know whether it was best for him to outflank the many strong points and anti-tank ditches. Stirling's role was the most dangerous: leading a small patrol he intended to penetrate the entire depth of the enemy lines to meet the First Army, which had landed in Algeria, and present its commanders with up-to-the-minute information about enemy strength and dispositions. It was a brave venture.[1]

On 23 January Stirling, with five Jeeps, crossed rough country and at dawn reached a road that would take him to Sousse. He needed petrol and planned to take it from the enemy. With no time to waste he tried a straightforward bluff, a tried and proved tactic. The SAS Jeeps, travelling at speed, passed through the middle of a German unit which was at breakfast. Some Germans looked curious, even surprised, but Stirling's men stared back at them with the same curiosity and drove on without a pause. They continued in this fashion for fully five hours, meeting one enemy unit after another. For petrol they milked unguarded enemy vehicles and took several full jerrycans.

The SAS Jeeps and their crews could have looked nothing like the orderly Germans or Italians. The vehicles and the men were scruffy, yet even their beards did not cause suspicion. Also, they were using Jeeps, a vehicle unknown in the Axis armies.

Suspicions were unsurprisingly aroused. The Germans were alarmed by Lieutenant Jordan's patrol and fired at it. Jordan shot up some enemy trucks and evaded capture but the presence of British troops in the area had been broadcast. Apparently the alarm had not reached the units which Stirling and his men had seen.

It was time to shelter for the day and Stirling led his patrol into the scrub of a hilly area. They camouflaged the Jeeps in a wadi, then spread out to find shade in which they could sleep off their exhaustion. It is said that Stirling failed to post sentries, though it may not have made much difference if he had done so. German troops located the Jeeps and, surrounding the area, began to search it with their usual Teutonic thoroughness. Stirling yelled to his men to scatter and get through if they could. Some did evade capture but Stirling and Private B. MacDermott hid in a cave and were captured. Two nights later Stirling, still unrecognised as the 'Phantom Major' – the name the Germans had given him – escaped but an Arab betrayed him and he was captured by Italians. The other missions, all planned by Stirling, were successful. In particular, Lieutenants Post and Jordan achieved great destruction, especially of enemy petrol and ammunition, though many of their men were captured.

With Stirling's capture the salad days of the SAS in North Africa were over. His enemies within the Army establishment tried to do away with the SAS and nearly succeeded when they incorporated it into general Army operations. Stirling's style of leadership did not help the surviving officers to follow where he had led. Impatient with record-keeping, administration and 'bumph', Stirling kept all information in his head. Nobody else had any real idea of what he planned for the future beyond his overall determination to 'play hell with the Hun'. Stirling's SAS had virtually no records beyond the papers which from time to time GHQ insisted on having.

If many of his superiors were relieved to see the back of Stirling, his captors – when they realised that they had in their hands the man who had caused them more trouble than an entire army – were delighted. Enemy officers came to look at him, as if he were a prize animal in a show. And indeed he was a prize. He was the greatest individual soldier of the war, with the possible exception of the German Otto Skorzeny, with whom he shared the same audacious streak. German officers did not understand how some British officers, Stirling most noticeably, were allowed to mount what were 'private' operations. They were not to know that for much of his active period Stirling had not been 'allowed'. He had taken the authority unto himself. It was not until the Tunisian manoeuvres that there was an 'official' role for the SAS.

Stirling's many friends, especially those close to him within the SAS, were appalled by his loss, all the more so because it need not have occurred. He was much too valuable an officer to have gone on the raid which was to be his last; he had numerous competent junior officers to whom he could have entrusted the mission. His apologists say that he badly wanted to see his brother in Algiers. If so, he could have flown there, and without asking permission.

David Stirling was the victim of his own success as a raider. Raiding was a heady, addictive experience and Stirling was hooked on it. Bluff had always worked for him and he had to try it yet again. Driving openly along roads being used by the enemy gave him an adrenaline satisfaction nothing else would have done. However, he was not alone in this; the SAS of North Africa and the campaigns to come were made up of such characters, always ready to dare the enemy to catch them.

The SAS patrols would have achieved even more had the Eighth Army planning staffs understood the potential for experienced and determined raiders. There is little evidence to suggest that the deskbound warriors in Cairo and Alexandria made any effort to follow SAS tactics. Time and again patrols were close to the enemy's lines of communications but were ordered not to proceed; sometimes they had no orders at all. Eventually, following Stirling's protests, Lieutenant-General Corbett, newly appointed as Chief-of-Staff to the Eighth Army, gave a direct order that HQ would establish a direct radio link with Stirling. Even then dedicated anti-Stirling obstructionists prevented any real cooperation from developing. It was little wonder that Stirling and his officers said sourly that they were fighting two enemies and British GHQ was the more dangerous.

Even after the war, I am sure that the military bureaucrats who wanted to see the SAS demolished would not have liked to be told that one SAS raid in North Africa destroyed more enemy aircraft in a day than the entire balloon barrage in Britain managed to bring down in the course of six years of war. They could not admit that Stirling was right.

SURVIVING AGAINST THE ODDS

Stories of endurance and fortitude of SAS men abound and some of them became widely known in the Army during the Second World War. One of the legendary figures of the North African campaign was Sergeant David Sillito, the navigator of a patrol sent to blow up the enemy coastal railway near Sidi Barrâni late in October 1942. The patrol was spotted by the enemy and attacked from the air and ground. The commander, Lieutenant Shorten, was killed, the transport was wrecked and the men scattered.

Sillito evaded his pursuers but he was alone, without food and, following the break-up of his patrol, he was equipped with nothing more than a waterbottle, a revolver and a compass. He could surrender, find a hideout until the Eighth Army vanguard caught up with him or, most difficult of all, he could try to rejoin his squadron. Sillito had been with the SAS for a relatively short time but he had heard many of the great survival exploits. However, nobody before him had set off to walk 200 miles into the featureless and harsh Great Sand Sea. The heat was intense so Sillito rested as best he could when the sun was high. With a few sticks he made his greatcoat into a low tent.

On day two his water ran out and he was forced to drink his urine, despite dire warnings in training that this could kill a man. Then urine ceased to flow. Still Sillito staggered on, suffering from blistered feet, acute sunburn, intermittent loss of memory and sand blindness. But he was not a navigator for nothing and on day eight he reached the northern edge of the Great Sand Sea. Here he came across some Jeep tracks, which he followed. He found a resting patrol from A Squadron SAS. Taken back to Kufra Oasis base, Sillito was well looked after by his mates and was ready for further duty within two weeks. Fortitude was commonplace in the SAS but Sillito's feat was considered exceptional and he was awarded the Military Medal. Later, during the Italian campaign, he was awarded another Military Medal for bravery.

Sergeant Sillito's adventure was by no means the longest ordeal in the history of British raiders. Lieutenant-Colonel Laycock and

Sergeant Terry had marched for forty-one days after the failed raid on the Rommelhaus (see Chapter 11), but 'Tanky' Challenor would well hold the unofficial record for endurance in terms of the length of time spent, the distance covered and illnesses survived.

In October 1943, with Lieutenant John Wedderburn, Challenor was sent to blow up the Bologna–Genoa railway. They chose their target well – a tunnel carrying a double line. They could spare only 6 lb of plastic explosive but in expert hands this was enough to deal with both lines. Wedderburn and Challenor had the satisfaction of seeing the up-train blow up and moments later the down-train was wrecked. The double explosion and the heavy locomotives being rammed into the walls caused the tunnel to collapse. Wedderburn and Challenor shook hands and hurried away from the scene of their destruction. A few days later they wrecked another train and then set off for British lines.

Challenor was suffering from malaria and hepatitis while Wedderburn's feet became blistered and torn. Winter had set in since they left base but sometimes they found shelter from the bitter cold with friendly Italian families who hated their own Fascists as much as they did the occupying Germans.

Wedderburn and Challenor separated at Christmastime 1943, each of them feeling that he was holding the other up. The officer was soon captured but Challenor, now really ill, pressed on before he too was captured. His captors accused him of being a spy and brutally beat him up but he aroused the sympathy of a woman who gave him a dress and in this garb he escaped. In the hills compassionate Italians cared for him as he fought malaria and pneumonia. Hidden by these friends at dangerous times, he stayed with them until the beginning of April, springtime, when he headed south again. Living rough and often starving, Challenor reached the German front lines where he was again arrested. Hoodwinking his captors, the intrepid Cockney again escaped and picked his way through the hazards of No-Man's-Land to reach British troops. It was seven months since he and Wedderburn had commenced their raid.

Much later, Tanky Challenor wrote a book about his adventures and in it he quotes himself on finally reaching safety and joining his mates: 'All I could say, over and over again, was, "I've done you, you bastards. I've done you, you bastards."' This comment in itself was in the best tradition of British raiders.

THE SAS'S 'SOFT REFRAIN', A RIPOSTE TO 'LILI MARLENE'

There is a song we always used to hear,
Out in the desert, romantic soft and clear.
Over the ether came the strain, that lilting refrain,
Each night again, of poor Lili Marlene, of poor Lili Marlene.

(Chorus)
Then back to Cairo we would steer,
And drink our beer with ne'er a tear,
And poor Lili Marlene's boyfriend will never see Marlene.

Check your ammunition, see your guns are right,
Wait until a convoy comes creeping through the night.
Then you can have some fun, my son,
And know the war is almost won,
And poor Marlene's boyfriend will never see Marlene.

Drive on to an airfield, thirty planes ahead,
Belching ammunition and filling them with lead
A flamer for you, a grave for Fritz,
Just like his planes, he's shot to bits,
And poor Marlene's boyfriend will never see Marlene.

Afrika Korps has sunk into the dust,
Gone are its Stukas, its tanks have turned to rust,
No more we'll hear the soft refrain,
That lilting strain it's night again.
And poor Marlene's boyfriend will never see Marlene.

flamer: gone up in flames and smoke
Stuka: the feared German dive bomber

Lili Marlene was a German barracks song. Translated into English, it became one of the most popular songs of British Army concert parties. The SAS's song was not as irresistibly catchy as '*Lili Marlene*' but it says something for the SAS's pitiless ferocity in action in North Africa.

1. In his memoirs Montgomery does not give Stirling a mention. The lack does the petulant Montgomery no credit. However, during one of several long talks I had with Montgomery in 1967, he referred to 'that splendid fellow David Stirling'.

5

THE LONG RANGE DESERT GROUP

SEEKING OUT THE ENEMY

In 1939, Major R.A. Bagnold of the Royal Corps of Signals, advocated the use of independent, long-distance patrols to observe enemies of the British in North Africa. Bagnold had explored much of North Africa before the war and his experience was needed in the area which was to become a great arena of conflict. He had been posted to East Africa but the troopship on which he was travelling suffered damage and docked in Alexandria. By chance his name came before General Sir Archibald Wavell, Commander-in-Chief Middle East, who could see that Bagnold was a rare specialist. No doubt Wavell communicated his professional interest in Bagnold to the staff of MEHQ but, like David Stirling and others, Bagnold found the more senior officers obstructive. When he said that trained reconnaissance patrols could provide much valuable information about enemy intentions and activities they professed not to believe him. People got lost in these oceans of sand, they said. Bagnold, a quiet man without Stirling's skill in self-publicity, said calmly that he could safely negotiate the Great Sand Sea. The colonels and major-generals above him said that was merely a dream.

On 10 June, Italy, with an empire in North Africa, entered the war as Nazi Germany's partner. This gave Bagnold a chance to claim that reconnaissance was urgent. On more conventional lines than those employed by Stirling, he secured an interview with his commander-in-chief and convinced that sagacious man his ideas were worth developing. Wavell authorised Bagnold to raise a unit that he proposed and, more than that, made it clear to the staff of MEHQ that this officer was to have everything that he asked for. Not a man to make enemies, Bagnold went quietly but determinedly about his mission.

He could draw his men from wherever he wished in the British,

Australian, Indian, Polish and New Zealand forces. Bagnold had precise ideas of the type of men he needed. He wanted supermen but of a special kind. They had to be exceptional in all the qualities that Bagnold sought – intelligence, resourcefulness, self-control, mental balance and toughness, adventurousness. He did not want death-or-glory boys or blazing extroverts. He knew that in the small patrols which he proposed the individuals had to be selfless and willing to compromise. Bagnold did not at first see his men as raiders but as scouts – the eyes and ears of the Army. Their job was to provide information.

Bagnold made no invidious comparisons among the nationalities available to him but the New Zealanders, who had arrived in the Middle East early in 1940, appealed to him. He also found much to admire in the Australians but he judged them to be more impatient than the New Zealanders – and in this he was correct. To get his hands on New Zealand recruits, Bagnold had to gain the approval of their leaders. He went straight to the Commander-in-Chief of the New Zealand Expeditionary Force, Major-General Sir Bernard Freyberg VC. Freyberg would not detail men for the new unit but he allowed them to volunteer. From the many Kiwis who did so, Bagnold chose 2 officers and 110 other ranks for his Long Range Desert Patrols, as the unit was first known before it became the Long Range Desert Group.

He spoke at some length to each man in order to assess his personality and intelligence, also to be frank with them about the dangerous demands of the job. No man wanted 'to think again'. The Kiwis did not disappoint Bagnold. They learnt quickly and since many of them were farmers they readily settled down to living outdoors in extremes of climate and the great diurnal range of temperatures. A large proportion of the New Zealanders were drivers, a bonus for Bagnold. Driving a 30 cwt heavily laden Chevrolet truck over vast areas of trackless sand and scrubland was rather different from road conditions in New Zealand but in this as in all things, the Kiwis were adaptable.

Equipment was in short supply and at first azimuth cards, chronometers and even theodolites were unprocurable. Bagnold had surveyed parts of his vast area before the war but large parts had never been charted, so the LRDG men did their own surveying and produced their own maps. The unit set up camp first at Abbassia Barracks, Cairo, and here the men did their initial training, some of it under one of the few Englishmen to join LRDG, Captain E. Mitford, a Tank Regiment officer who before the war had visited Kufra Oasis, a genuine claim to fame in 1940.

Bagnold obtained some of the best radios in the world, sets with a range of 1,200 miles. He taught his wireless operators to lavish as much care on their equipment as they would their own baby and they developed great skill. In a way, the success of the LRDG was in their hands; if reconnaissance Intelligence did not reach the right officers at the right time then the LRDG had failed. Similarly, the navigators had immense responsibility. Patrols could perish if they were not efficient.

Between the time that LRDG commenced active operations in August 1940 and the end of the desert war in March 1943 at least one patrol was always on duty. The Army at large saw little of the LRDG men but whenever they did appear in 'public' they looked like romantic buccaneers in their Arab headdress, worn for comfort and coolness, and open sandals. In the cold they wore sheepskins, some of which were made into coats and cloaks in New Zealand. As with the SAS, there was little military formality in dress or behaviour.

Every man could drive, just as every member of LRDG became proficient in the use of an array of weapons, including the 37 mm Bofors anti-aircraft gun. Bagnold was promoted to lieutenant-colonel and because of early successes LRDG was increased in size to two squadrons each of three patrols. Bagnold needed additional men but, being prevented from obtaining more New Zealanders, he sought them from the Southern Rhodesian Brigade, from Yeomanry units and elsewhere. Two officers and thirty-six men from Guards regiments became G Patrol. This group and the New Zealander's T Patrol did much work together.

It was a tough life. A patrol was out from base for three weeks and it covered up to 2,000 miles. Since a Chev under desert conditions gave no more than six miles per gallon, the trucks had to carry large quantities of fuel. The water ration was six pints per man per day which, during the hottest months, was barely adequate. Sometimes ten or even more trucks set off on patrol simply in order to carry petrol, water, ammunition and other supplies.

Like the SAS, LRDG patrols raided airfields and attacked isolated Italian forts. An early success was the raid on Murzuk, near the Chad border, during which three Italian aircraft and a fuel dump were destroyed. Murzuk was so far away from any of the North African campaigning areas that the startled Italians could not even imagine where the raiders had come from. It was unbelievable that they could cross thousands of miles of sand sea. The LRDG's ability to raid at will destroyed the Italians' morale.

Early in March 1941 General Leclerc, commanding the Free

French in North Africa, captured Kufra Oasis from the Italians, thus providing the LRDG and SAS patrols with their main forward base. In desert terms, Kufra was a desirable place if only because it had limitless supplies of fresh water as well as large areas of palm trees, which the raiders sought the moment they arrived in Kufra from a mission. After exposure day after day to the blistering metallic sun it was an immense relief to be in deep shade. Patrol commanders tried to give their men three days' rest in Kufra.

It is difficult to single out one exploit as outstanding from among the hundreds of missions carried out by the LRDG. But without claiming it to be 'the best', I can perhaps put forward a classic patrol by the LRDG and one which showed opportunism and audacity. Late in 1941 the LRDG's G1 Patrol, under Captain A.M. Hay, emerged from the desert wastes to find targets on the main Axis communications artery, the road from Benghazi eastwards to the fighting areas. The men already knew that the enemy base at Beda Fomm was a fruitful operating area. An Italian pilot spotted them and dropped bombs but the patrol's Bren guns frightened him off. The Italian called up a German Junkers 87, whose crew made strafing and machine-gun and bombing runs against the LRDG trucks. Remarkably, Hay, his men and their vehicles came through this ordeal unscathed. They headed for concealment in a wadi they knew of old and here they spent an uneventful night. Scouting on foot next day, Hay lay on a low ridge and swept the country through his binoculars. He located a choice of targets but the best was a vehicle park close to some huts built as a form of way-station for crews on their long journeys to and from the front. That evening at dusk Captain Hay led his trucks at a steady speed along the Benghazi road. Enemy transport was moving in both directions but nobody paid any attention to the 30 cwt trucks, perhaps because the Germans and Italians both used captured British vehicles and they were accustomed to seeing Chevs. In any case, they would never have thought that the British would drive openly along the highway.

At the way-station recreation huts Hay's drivers turned off the road and unhurriedly took up the best possible positions for attack. Then they began a raking fire with Vickers machine-guns, Bren machine-guns, tank-attack rifles and Tommy-guns. Enemy soldiers, shocked by the surprise of the attack, fell in heaps and some of their vehicles caught fire. To add to the confusion the LRDG troopers tossed grenades into the huts. By the time the enemy troops reorganised themselves the raiders had long gone. Enemy aircraft were soon hunting for them and their search continued throughout the following day. Again well camouflaged in a friendly wadi the patrol slept without an alarm. When dusk once more gave him cover, Hay machine-gunned and set fire to a petrol tanker. Then, as ordered, he returned to base in Siwa Oasis without having suffered a single casualty. His raid had been as successful as any carried out by the SAS.

Such raids were commonplace. Even when certain teams were resting from road watch or convoy attack they asked permission to make raids. Sometimes they told the CO that they needed the 'practice', on other occasions they complained that they were bored with oasis resting. In April 1942, two New Zealand patrols seeking relief from boredom ambushed a convoy on the Benghazi–Agedabia road; they destroyed all the vehicles and killed many enemy soldiers. Since there was unlikely to be a hunt for some time the Kiwis drove down the road to where drivers of another convoy had abandoned their vehicles in a panic after hearing the earlier shooting. The Kiwis souvenired some useful supplies from these trucks and then destroyed them. Nothing moved along that road for nearly 30 hours and then it was convoyed by armoured cars. In yet another raid, the New Zealanders destroyed twenty-four aircraft on an airfield and damaged another twelve. This was a raiding coup that drew admiration from Paddy Mayne, the SAS's king of airfield attacks. The New Zealanders have been colourfully described as 'the richest marrow in the LRDG's veins'. It is an appealing metaphor.[1]

When the campaigns in North Africa came to an end the LRDG was only briefly unemployed. It operated first in Italy, then in Yugoslavia, Greece, Albania and the Dalmatian Islands in the Adriatic. One of its most outstanding successes in Europe took place in northern Greece in the autumn of 1944. Major Stormonth-Darling, Captain W.L. Armstrong and two patrols parachuted into Florina with orders to harass the German forces as they withdrew northwards. There was no desert here, no really long-range activity but harassment was now the LRDG's principal *raison d'être*. Brilliant at the art of ambush setting, Stormonth-Darling found just the spot he needed – on a bend in a road much used by the enemy, with scrub nearby for concealment. Captain Armstrong, an Army engineer with the sappers included in the party, mined two road culverts during darkness. Stormonth-Darling gave his men explicit orders: 'No individual firing. Total lack of movement. No smoking. Fire only when the mines are exploded and then choose your targets.'

He arranged his men around the killing site so that the Germans would be trapped no matter which direction they took. The major and captain agreed that they did not want to bag only a single truck; sooner or later a really good target would present itself. That target turned out to be four vehicles travelling much too close together for their own safety. Armstrong fired the charges. The two culverts, the road and chunks of the trucks went up in the great explosion. The road was completely blocked and as the surviving Germans ran from their wrecked vehicles they met a hail of fire from an enemy they never saw.

The LRDG men left fifty dead Germans on the scene and disappeared into the hills without loss to themselves. How many Germans were wounded? None, because the LRDG's small-arms fire was so devastating that every enemy soldier was hit several times. The tally of fifty enemy dead for a single raid was by no means a record. A raiding party under Captain S.N. Eastwood, in Albania killed eighty enemy at a roadblock and destroyed a tank and several trucks.

By this time Bagnold had long since been promoted to colonel and transferred to a staff appointment in Cairo but command of LRDG passed to an equally capable officer, Lieutenant-Colonel Guy Prendergast.

LRDG operations were not always successful, mainly because the men were asked to perform work for which their training and equipment were unsuitable. On one sad occasion LRDG lost forty men, an almost catastrophic blow to a small unit. In November 1943 on the island of Leros, fighting the Germans, LRDG had more than fifty men taken prisoner. Killed in action was Jake Easonsmith, who had succeeded Prendergast as the LRDG CO.

Always reinforced with hand-picked men, LRDG's services were in constant demand. Patrols frequently operated at a long distance from their HQ and for some periods they were in action in three different countries at the one time. This called for outstanding command by junior officers and NCOs and for a high standard of performance by all ranks. Some missions were purely for reconnaissance, like those of the early days in the desert, others were raids. No two jobs were ever the same and LRDG produced some of the most versatile warriors of the entire war.

As the war ended in Europe, LRDG, through its CO, Lieutenant-Colonel David Lloyd Owen, volunteered to serve in South-East Asia. This was a generous offer because most of the men were entitled to demobilisation. They made only one request: they wanted to serve as a unit. The War Office rejected their offer and in August 1945 LRDG was disbanded.

LRDG had been on active service for five years and a handful of its original members remained until the end. It was one of the most successful independent outfits of the war. No prima donnas, these quiet and unassuming raiders came out of the great conflict with an unblemished reputation which deserved more public – and official – recognition than LRDG ever received.

1.   William Seymour in *British Special Forces*, Sidgwick & Jackson, 1985.

6

GLIDER-BORNE RAIDERS AT EBEN EMAEL, MAY 1940

In the 1930s when the Belgian government and Army built Fort Emael to cover a vital part of their frontier with Germany they depended on the expertise of two German construction companies to do the job – Hochtief AG Essen and Dycherhoff & Widmann of Wiesbaden. These foreign contractors did their work splendidly, creating one of the most powerful fortresses in history. It was positioned 60 metres above the Albert Canal, a significant height in this flat land; it covered 900 metres from north to south and its arrowhead formation pointed towards Maastricht and Germany. The whole was protected by a clever interlocking of anti-tank ditches and water-filled moats.

Eben Emael's armaments were impressive. Casemates or cupolas housed revolving guns and anti-aircraft artillery. In the fort's southern area were three triple and two twin 75 mm gun batteries and one heavy twin 120 mm long-range battery. Ground defence for the fort itself was multiple anti-tank guns in blockhouses. The outer defences also contained gun positions. Machine-guns proliferated, so sited that they could sweep every angle of approach by enemy infantry; in addition, they protected the cupolas.

Virtually nothing in Eben Emael was exposed. Most troop accommodation and all ammunition chambers and control centres were underground. In addition, under normal circumstances it would not be necessary for the 1,200-strong garrison to move about on the surface; they used the 7 kilometres of tunnels, which linked the various parts of the amazing complex.

The Germans, the only possible enemy for the Belgians, had one great advantage. The German General Staff, courtesy of Hochtief and Dycherhoff & Widmann, had a copy of Eben Emael's blueprints down to the last rivet in the heavy steel armoured sheets. The Belgians had been too trusting, especially as their own engineers or

French experts could have built the fort. Still, the Belgians were aware of Germany's territorial intentions. They had been well advertised during the First World War, when vast German armies violated Belgian neutrality and swept through the small country to attack France. Fort Eben Emael and other forts covering the River Meuse (Maas) were supposed to prevent a second such invasion. When war broke out in September 1939 and Hitler's blitzkrieg swept through Poland and Scandinavia it was obvious that the German dictator, his Nazi colleagues and the German High Command would be considering their next step. However, they were not entirely convinced that their powerful Panzer forces could so readily breach France's immensely powerful Maginot Line. The solution was simple – they could flow around the northern end of these strong fortifications. This meant breaking Belgian's defences and in particular Eben Emael.

It seems to have been Hitler himself who made what was then a revolutionary suggestion. Why not land gliders carrying a highly trained raiding party within the central defences by night and thus take the defenders by surprise? He put this idea to General Kurt Student, commander of the 7th Airborne Division, at the Berlin Chancellory on 27 October 1939.

Student asked for a day to consider and returned to say that the job could be done. He wanted only one change from the plan proposed by Hitler and his field marshals; the attack should be made in enough daylight for the glider pilots and the assault troops to see their targets. Otherwise, Student warned, some gliders would crash or miss the landing area altogether. He got his way.

Using the building blueprints, German engineers built a replica of Fort Eben Emael, near Hildesheim, and a cordon was thrown around it to keep out curious people and to keep in the raiding party, which was to practise for six months. As with so many other successful raids, the Eben Emael operation proved the need for detailed planning. The Germans were nothing if not methodical. General Student named Captain Erich Koch to command the raid, code-named Assault Force Granite. Koch, given a free hand to choose his men, selected paratrooper Lieutenant Rudolf Witzig to lead the assault. The men were hand-picked from the 7th Flieger Division. The entire 'force' numbered eighty-five.

There was nothing new about military gliders in Germany. Forbidden by the Treaty of Versailles to create an air force, the Germans evaded this restriction by building gliders, which in the 1920s and 1930s were not seen by the British and French as threatening. Nobody from outside queried the size of the gliders in the 1930s but the DFS 230 could carry 1½ tons. The towing aircraft was the Junkers 52, which, in the 1930s, looked like an innocent commercial transport.

Assault Force Granite troops would use their normal automatic rifles and grenades but their explosives were anything but normal. They were of the hollow-shape type, proven successful in experiments but not previously used in combat. In essence, the pack consisted of a cylindrical cavity around which Monobel explosive was tightly packed. The whole pack was fixed against the target, be it tank or fortress or an artillery piece. When set off, the explosion was channelled into the cavity of hollow under tremendous pressure and high temperature. The damage was done by a thin jet of molten steel which penetrated the steel, causing terrible destruction internally. Great chunks of the target tank or fort armour sprayed around the tank's cabin or the gun casemate. Nobody could survive and the enemy target was wrecked. The hollow-shape worked with equal devastation on concrete. Assault Force Granite mostly used a one-man, 25 lb charge but a few monsters of 100 lb were employed.

Take-off for the raid was from Ostheim airfield at 3.35 a.m. on 9 May 1940. Not until the men were filing aboard their gliders did they know of their target, but where it was did not concern them. From their rehearsals they knew exactly what they were approaching. According to German records, the raiders sang battle songs as the Junkers and their towed gliders became airborne in good flying weather. Nothing had been forgotten – even down to the line of beacons which guided the pilots on their way from the Rhine to the Belgian border. Witzig and his team flew in the last glider.

The operation could have been wrecked in the first few minutes. Witzig's glider pilot had to take violent evasive action to avoid collision with a Junkers that crossed between his glider and the mother plane. The tow rope broke but the skilled pilot set down the glider in a field west of the Rhine. Assault Force Granite was without its commander. Witzig did not panic. Ordering his men to clear the field of obstructions he hunted for a telephone in order to call for another tow.

Major Jean Jottrand, CO of Eben Emael, was asleep in his underground room, when a telephone call from the duty officer awoke him. Army HQ in Liège was alerting Jottrand to reports that German troop movements had been observed and he should come to stand by. Jottrand sounded the alarm and submarine-type hooters brought his men running to their posts – though about half the garrison were asleep in their billets in nearby villages.

The next disturbing reports were that German paras were landing at the Meuse bridges of Vroenhoven, Canne and Veldwezelt. Anxiously, Jottrand and his men peered through their observation slits, which were covered with reinforced glass. A few spotters with binoculars were above ground but with dawn just breaking their glasses were not much use. Within minutes binoculars were not needed as strange shapes came rushing silently at the fort. No one there had ever seen gliders: they were very strange and menacing.

Jottrand kept his head. Telephoning the officers in charge of the three bridges, he ordered them to blow the demolition charges. One officer managed to do this just as the first paras landed but the other bridge commanders were too slow to act and the German paras captured the river crossings intact. Jottrand also ordered his anti-aircraft gunners to open fire. Bullets hit one glider but the Belgian fire had come too late. The gliders were down and German soldiers were running purposefully to their targets.

They fixed hollow-shape charges to the big guns and cupolas, to the heavy steel which covered parts of the fort and to the ventilation ducts. Eben Emael was now a death trap and some of the garrison were crushed and dismembered by tons of jagged flying steel. Others were burnt to death; many were choking from fumes. All the men who had survived the initial assault were in shock. Some Belgian soldiers recovered and fought bravely but the Germans were elusive and the defenders were handicapped by not knowing from which direction the next attack might come.

The raiders had been so well trained that Lieutenant Witzig's absence was hardly noticed. Master Sergeant Wenzel, the senior NCO, took over command and was doing his job efficiently when he was surprised to see a lone glider rush to a screeching stop on an open part of the inner fort. From it stepped Witzig who calmly thanked Wenzel and relieved him of the command. He had gone to enormous trouble to get a replacement Junkers and under normal circumstances he would have been reprimanded for the arrogant insolence he had displayed in order to commandeer the Junkers.

Witzig was now needed. German sappers, not directly part of the glider-borne team, were in combat on the far side of the fort and needed direction. Belgian guns above the fort had come into action and fired on Witzig and his men. They sheltered inside the wrecked cupolas but Witzig's force could not possibly beat off any strong Belgian counter-attack. However, an attack in such strength did not eventuate. Major Jottrand rallied his survivors, who made small-scale and hopeless charges. They held out during the night but Witzig saw no point in risking his men by making attacks down the tunnels of

Eben Emael. By now the German Panzer columns, free from the menace of Eben Emael, were bypassing the fort and flooding into Belgium, and soon into France.

Witzig pinpointed the position of Major Jottrand's command post; a hollow-shape charge destroyed it. Even then, pockets of Belgians remained holed-up in the depths of Eben Emael. Witzig waited for them to come out and surrender.

It was ascertained that a mere 55 raiders of Assault Force Granite fought at Eben Emael. The others of the original 85 had simply not arrived, mostly because their gliders overshot the small target area. Six Germans were killed and 15 wounded. The Belgians lost 23 dead and 59 wounded, with hundreds taken prisoner.

The raiders' overwhelming success, brought about by new methods of attack, was a great boost to German Army morale. Not that it needed it – the blitzkrieg had already convinced them that they were unbeatable. The victory using gliders at the fort and paras at the bridges also influenced Hitler and his generals to employ similar tactics elsewhere, notably on Crete the following year. The Germans captured Crete but they suffered colossal losses among paras and glider-borne troops at the hands of disciplined veteran New Zealanders and Australians. The Eben Emael operation had worked because it was conceived and carried out by a highly motivated and superbly trained small number of men and because it was a genuine surprise attack.

Lieutenant Witzig saw much more action during the war and won promotion to colonel. Captain Koch, who had evolved the tactics and directed the training for the raid, was killed in a wartime car accident.

The Allied High Command believed that the Germans would fight fiercely to hold Eben Emael against the advancing American troops in 1945. But the Germans had proved the fort to be indefensible and having used it as a repair workshop for guns, tanks and vehicles, they abandoned it. The American troops captured an empty shell.

I wonder if Koch and Witzig ever thanked Hochtief AG Essen and Dycherhoff & Widmann for their considerable help in the raiding coup of 1940. There is a lesson in Eben Emael for nations which build forts, or for that matter tanks, guns, warships and aircraft. Employ your own people as contractors and make sure that the blueprints never leave the premises!

7

COMMANDOS IN THE ARCTIC, 1941

LOFOTEN, OPERATION CLAYMORE, MARCH 1941. VAAGSO, OPERATION ARCHERY, DECEMBER 1941

These two raids, both in the cold North, were made in 1941 when the Commando idea was still being tested. Lofoten, the smaller raid, was a total success. Vaagso turned into a bloody battle. Lofoten proved that raids could be effective. Vaagso was the first attempt to attack a well-defended target exploiting the element of surprise. In a way, Lofoten inspired Vaagso, which some officers who took part considered the perfect raid. It was not quite that but it did show the amount of damage that could be done to the enemy in a few short hours. The two accounts need to be read together.

FIRST LOFOTEN RAID

In autumn 1940 great things were expected from the new Commando force of the Special Service Brigade under Brigadier J.C. Haydon. He was as disappointed as his men when one scheme after another fell through.

The first was to have been the invasion and capture of the Azores and for this operation five of the Commando units and many assault vessels were concentrated at Inveraray, Scotland. Many assault exercises were carried out but the operation was cancelled, for no good reason.

Admiral of the Fleet Sir Roger Keyes, Director of Combined Operations, had been as frustrated as anybody else by what he called 'the negative power which controls the war machine in Whitehall'. He now chose Pantellaria in the Mediterranean as his next objective and this time concentrated his force on the Isle of Arran, where further training and practice assaults got under way. On orders from 'on high' this operation was also cancelled. The troops, whose morale was suffering, naturally enough blamed

Haydon, the most senior officer they saw. A vigorous man, Haydon revived the men's spirits with optimistic talks and by encouraging his officers to do a real job of leadership by inventive and enterprising training as well as setting an example of enthusiasm. Part of his SSB was sent to the Middle East as 'Layforce', and this evidence of action cheered the men. The Commando units were also reorganised but best of all a new raid was planned and this time there was every sign that it would be carried out.

The objectives were the German-occupied Lofoten Islands, off the coast of Norway. Here the Commandos were to destroy the fish-oil factories, which would deprive the Germans of glycerine for the manufacture of explosives. In addition, the raiders were to sink enemy shipping, to enlist volunteers for the Norwegian Army and Navy units in Britain and to capture supporters of the traitor Vidkun Quisling, whom the Norwegian government-in-exile wanted to punish.

On 21 February 1941 the Commandos boarded the Landing Ships Infantry (LSI), *Queen Emma* and *Princess Beatrix*, and were greatly heartened. They spent a week at Scapa Flow in final training and administration. Then the ports of Stamsund and Henningsvaer were apportioned to No. 3 Commando, Solvaer and Brettesnes to No. 4 Commando. Detachments of engineers and Norwegian soldiers were attached to each Commando, for demolition and interpreting tasks respectively. The force was under Brigadier Haydon, while Lieutenant-Colonel John Durnford-Slater of No. 3 was the principal officer leading the assault. Each Commando numbered 250 men, with 52 men of the Royal Engineers and 52 from the Norwegian units.

That this Commando action was taken seriously became evident from the powerful naval support, the battleships HMS *Nelson*, HMS *King George V*, and HMS *Nigeria*, together with five destroyers. Throughout the war few Commando operations of this size had such a strong naval presence.

At Stamsund the Commandos were genuinely disappointed to find no opposition on land, though the armed trawler *Krebs* gamely gave battle to HMS *Somali*. It was set on fire and the crew surrendered. As with all Commando raids made in force one of the main purposes was destruction and here the troops achieved their goal. They sank 11 ships – a total of 20,000 tons – while one trawler was manned and taken back to Britain. About 20 factories were demolished and 800,000 gallons of oil and petrol were fired. No. 4 Commando captured 216 Germans, 66 Quislings were arrested and 315 Norwegian volunteers, including one woman, were taken off.

The Commando spirit of enterprise was shown in an unusual

way at Lofoten. Each soldier had been issued with a 'float' of 100 kroner, should he be left behind and have to make his way back to Britain. This money was collected from the men when they returned to their ships but one sergeant was 30 kroner short. It turned out that he had paid this money to a curvaceous Norwegian blonde for 'services rendered'. He never lived down the nickname he was given that day – 'Thirty-kroner Smith'. I have not used his real name!

The only British casualty on the raid was an officer who shot himself in the foot with his Colt automatic, which he had pushed into his trouser pocket. When he went to take it out his finger pressed the trigger.

A Department of Information film team made a record of the Lofoten raid and, at a time when the British were having few successes, it proved to be an effective piece of propaganda. The footage showing captured German prisoners and Quisling traitors being arrested was especially heartening and was shown in cinemas across Britain.

Lofoten might not have been a dramatic raid but it was certainly good experience for the planners, the officers and the men themselves. The Commando idea worked.

VAAGSO

The strategic objective for the Vaagso operation deep in a Norwegian fjord was to keep German troops there and deter Hitler's generals from sending them to Europe and Africa. The tactical objectives were the German-held installations, mainly in South Vaagso, where it was believed 150 Germans and a tank were stationed.

Vaagso was a tough nut. The Germans had mounted a battery of captured Belgian 75 mm guns on the island of Maaloy, in the middle of Ulvesund, thus blocking access into Vaagso fjord. On Rugunso Island, southern Vaagso, they had four more guns, while a mobile battery of 105 mm howitzers at Halsor blocked the northern entrance into Ulvesund. Air reconnaissance showed that the Luftwaffe had three airfields within range, at Stavangar, Herdla and Trondheim. This was bad news but a report that the Norwegian resistance had spotted no heavy enemy warships in the area was of some comfort.

The raiding force consisted of 51 officers and 525 men. No. 3 Commando provided five assault parties, strengthened by two platoons from No. 2 Commando. Attached were some Royal Engineers and a party of Norwegians. In command was Lieutenant-Colonel John Durnford-Slater, who had been at Lofoten. His air support was considerable. Ten Hampden bombers would blast the defences and smother them with canister smoke, to help the raiders to land more safely. Blenheims and Beaufighters would keep the Luftwaffe squadrons away from the action, it was hoped. The naval force was also reasonably strong, with the light cruiser HMS *Kenya*, four destroyers, a submarine and two infantry assault ships, HMS *Prince Charles* and HMS *Prince Leopold*. These two had been purpose-built; no longer did landing parties have to depend on converted holiday ferries.

Brigadier John Haydon, the Commando chief, was in command over Durnford-Slater while Rear-Admiral Burroughs was in charge of the small fleet. For better coordination in the attack, Haydon sailed in Burroughs's flagship, *Kenya*.

As for the Lofoten operation, the assembly point was Scapa Flow for final exercises and study of the latest reconnaissance photos and reports. On Christmas Eve the force for Operation Archery sailed for the Shetlands and ran into a gale so severe that ships were damaged. To repair them in Sullom Voe and to give the seriously seasick men time to recover, Admiral Burroughs postponed the raid for 24 hours.

The planners had deliberately decided on midwinter for the raid because at this time in the northern latitudes darkness prevailed for most of the time, relieved only by a few hours of twilight. In line-ahead, the ships steamed unnoticed into Vaagso fjord on 27 December and as they turned north the battle ensign was hoisted and landing craft were lowered into the icy water. The time was 8.40 a.m. HMS *Kenya* opened the naval bombardment at 8.40, its star shells lighting up Maaloy Island. Within 10 minutes the warships had fired 500 6-inch shells into a space of 250 square yards.

The Germans were taken by surprise, and their response was made worse by incompetence. A signaller on duty at the naval signals station on Maaloy received a message flashed by lamp, informing him of the British attack. Unable to get through by telephone – the soldier whose responsibility it was to answer the phone was cleaning his officer's boots next to the set – the signaller ran to a boat and rowed furiously to the office of the German naval commandant on Vaagso to deliver the warning. 'Did you warn the army gunners on Maaloy?' he was asked. 'No sir,' he answered. 'That is a military battery and this is a naval message.'

The Commandos, who had expected Maaloy to be the most difficult of their targets, overran the island in 8 minutes. First ashore was Captain Peter Young, commanding 6 Troop of No. 3 Commando. The only fighting took place at the German Command

Port, where Captain Martin Linge of the Norwegian Army, a highly respected Commando officer, was killed.

At Mortenes, on the other side of Ulvesund, a herring-oil factory was destroyed. However, South Vaagso was posing problems. To make it worse, a Hampden bomber, hit by anti-aircraft fire, released an incendiary bomb just before it crashed. It fell onto a landing craft, blowing up a stock of ammunition and grenades and causing many casualties among the Commandos.

The Germans defended every house, supported by snipers in concealed positions on a nearby hillside. At a critical moment Peter Young arrived with eighteen men from Maaloy Island, together with the floating reserve despatched by Brigadier Haydon. Germans in two warehouses blocked the Commandos' advance but Young rushed one of them under covering fire. Some of his men reached it unscathed and threw grenades inside. This did not end the enemy resistance and they also threw grenades. Lieutenant O'Flaherty and Trooper Sherington, armed with machine-guns, dashed through the front door, followed closely by Peter Young. He saw the lieutenant and trooper fall but they managed to pull themselves up and stagger out, O'Flaherty wounded in the face and Sherington in the leg. Young had the answer to this impasse; he sent for a bucket of petrol which was flung inside and fired. The building went up in flames, the Germans with it.

While all this was going on, Lieutenant-Colonel Durnford-Slater had been unable to get reports because his radios had broken down. He could hear fighting all over the place and to find out what was going on he marched boldly down the main street, accompanied by his signals officer and some runners. An enemy soldier threw a grenade at him and it exploded between his feet. He was untouched but a runner was badly wounded. The colonel was renowned for his luck and it had kept him alive once more.

The Arctic day was ending. Having gained control of his scattered men, Durnford-Slater fought several engagements and had achieved most of the initial objectives by nightfall. He ordered a withdrawal and under protection from the rearguard, the men backed away through the flames, smoke and gunfire, making for the rendezvous point.

The Navy had carried out its part of the raid. *Kenya* knocked out the battery of captured Belgian guns and put up such a fury of anti-aircraft fire that most of the Luftwaffe bombers which had appeared over the fjord were driven off. The destroyers sank two armed trawlers, an armed tug, a small coaster and four cargo boats, a total of 16,000 tons of shipping. The Navy lost two ratings killed and two officers and a rating wounded. The RAF bombers attacked Maaloy Island, as arranged, and laid down a smokescreen. Protected by the Beaufighters and Blenheims, the Hampdens destroyed the wooden runways of the German airfield at Herdla. Eight British planes were lost.

The Commandos had 2 officers and 20 men killed and 57 wounded; most casualties occurred as a result of the incendiary falling onto the landing barge. More than 120 Germans were killed and about the same number wounded. Ninety-eight German prisoners were taken back to Britain. Eighty Norwegians from Vaagso volunteered to leave with the Commandos and join the Free Norwegian forces in Britain. Four Quislings were not asked if they volunteered to leave. Two Norwegian girls, volunteer 'comfort girls' for the Germans, were taken aboard *Prince Charles* and locked in the medical officer's cabin. They dressed in the MO's pyjamas and tried to flirt with the sentry at the door. Taken back to Britain, they were handed over to the Norwegian authorities and sent to prison. The Norwegian government-in-exile, dealt severely with Quislings and collaborators.

Operation Archery had been a success and demonstrated that with adequate planning and training for specific tasks Commandos could do a lot on the periphery of the war. In 1941 there was no battle in Europe. The Commandos and other special forces provided the only ground retaliation against the Germans.

As a diversionary tactic, the Lofoten Islands had been attacked in Operation Anklet while the Vaagso operation was in progress. Anklet was commanded by Lieutenant-Colonel S.S. Harrison of No. 12 Commando. In their new winter uniform of white overalls and hoods, they remained on Lofoten for two days, destroying any installations the Germans had replaced since Operation Claymore in March. They took with them back to England all the Germans on Lofoten. The Commandos handed over to the islanders what was left of the garrison's Christmas rations. It was said among the Commandos that the crews of the two German radio stations on Lofoten surrendered to Lance-Corporal Webb, who demonstrated once again Commando nerve and bluff. Alone, Webb burst in upon the well-armed Germans and shouted 'Achtung! Schpitfuer!' They were the only German words he knew but they terrified the Germans. (Attention! Spitfire!)

The Commandos kept a tally of the damage they had caused to German property at Vaagso: 1 15-ton tank, trucks, 5 guns, numerous petrol storage tanks, 3 ammunition stores, barracks and telephone installations, several hundred beach mines, 'assorted' searchlights, 3 lighthouses, 4 factories, enemy offices.

While all this was helpful to the British war effort and damaging to the Germans, there never was the slightest hope that Hitler and his generals would reinforce Vaagso with so many men that their absence from North Africa and Europe would cause the German command any difficulties. This had been an impossibly ambitious strategic objective. Vaagso was just not important enough to the Germans to strengthen with a bigger garrison. In fact, having repaired the damage, they then kept a smaller garrison there.

When No. 3 Commando returned to base after Vaagso, Durnford-Slater addressed his men:

Future operations must be regarded as the highlights of our lives. A very few of you didn't like the Vaagso operation and must leave the unit forthwith. Your behaviour and turn-out must be irreproachable at all times. Have all the fun that's going – drink, gambling, chasing the girls and so on – if it appeals to you. But if these things interfere with your work they must be put aside. Personally, I have long quiet periods without any of these diversions and recommend you to do the same. You must always behave and look like super soldiers. If you cannot, then there is no place for you in No. 3 Commando.

He was asking for the impossible – but he got it.

8

BUCCANEERS OF THE SBS

The letters SBS are generally taken to mean Special Boat Service, in the way that the SAS was the Special Air Service, but the first unit of water-borne raiders was the Folboat Troop, later it became the Special Boat Section. Part of it was designated Z SBS and later still the Special Boat Squadron was the proper title for the unit. Its appellation made no difference to the type of work it did and fifty years after the war the acronyms are unimportant, except for clinical historical records. In general, in this account I take SBS to mean the popular Special Boat Service.

The SAS, LRDG, SBS and all the élite groups, including the first Commando units, attracted officers of high calibre. It must also be admitted that they attracted some officers who, while willing enough, did not possess the nerve required for independent missions in which junior officers had to make decisions that would normally be the responsibility of more senior officers. Sergeants too carried responsibilities that in conventional units would belong to officers.

Other men volunteered for duty with the raiders for quite the wrong reason: they believed that life in these special outfits was 'glamorous' and that when they returned after a mission to their base, such as Cairo, they would be socially lionised. They were soon spotted by raider leaders as being unsuited for life and work in an élite unit and their papers were marked with the terminal RTU – Return to Unit.

Scots and Irishmen, the latter predominantly from Eire, made up 60 per cent of the SBS's rank and file. It was always said by their leaders that men of these nationalities made good raiders because they actually liked to fight and to take risks. They were known to be difficult to discipline but not by SBS (or SAS) officers, who led from the front and by example and who did not impose traditional parade ground discipline.

One officer who would obviously make the grade right from the start was Roger Courtney, big, bold and young. Before the war he had been a white hunter in Africa, leading parties on safari. For sheer adventure he had canoed down the Nile from Lake Victoria.

When war broke out he was an officer in the Palestinian police. He joined the Commandos as soon as he heard about their existence and because of his confidence, leadership and experience in rough country he was commissioned and posted to the Combined Training Centre. With an innovative and enterprising mind similar to that of David Stirling, he submitted a paper suggesting that collapsible Folboats could be employed to destroy any enemy shipping. His superiors ridiculed the idea. Courtney asserted that he could board a naval vessel, plant bombs and get away again. He set about proving himself right. His chosen target was HMS *Glengyle*, in the Clyde river. The crew was warned of Courtney's intention in order to make *Glengyle* as secure as an enemy ship with sentries would be but Courtney welcomed the realism. He swam out to the *Glengyle*, removed the cover from a pom-pom gun and departed without being seen. In his dripping wet trunks he pushed his way into a high-level conference, attended by *Glengyle*'s captain, and presented him with the pom-pom cover.

Some of Courtney's seniors resented his 'impertinence', but when Admiral Sir Roger Keyes, Chief of Combined Operations, heard of the exploit he preferred to describe it as audacious. Intrigued, he asked Courtney to make a similar 'attack' on a submarine depot ship. Again Courtney took evidence from the ship to prove that he had been there and could have placed mines in or under it.

In July 1940 Keyes authorised Courtney to form the very first 'special' boat section, though it was not then known as such. Officially it was the Folboat Troop. His ideas, developed during his training of canoeists for the Folboats, became standard for all Folboats. They operated from submarines, making reconnaissances along shores and estuaries and carrying out sabotage.

Promoted to major in 1941 Courtney was ordered to form a second Folboat section. As always, he imbued his men with a sense of mission and a dedication to their work. At times there were suggestions that naval personnel would be more natural material for Folboat crews but Courtney successfully resisted this. Anybody could paddle a canoe, he argued, but only highly trained soldiers would have the skills needed for his type of land warfare. The Folboats were merely a means to an end; the targets were mostly on land or moored to land.

One of his groups was based on Malta and devoted much of its time to train-wrecking in Italy and Sicily. His Alexandria group chose targets along the Egyptian and Libyan coasts.

The SBS men acquired so much military knowledge that they became valuable commodities and were deployed elsewhere.

Courtney himself left the Mediterranean theatre in December 1941. By then or soon after SBS was a relatively large organisation. Courtney would have been disappointed in November 1942 when it was attached to the SAS but this change of management in no way cramped the Folboat warriors' style and they continued to call themselves SBS.

An ambitious Folboat operation was carried out in Boulogne harbour early in April 1942 by Captain G.S.C. Montanaro, who had been Courtney's deputy, and Trooper J.G. Preece.

An MTB took the two raiders into the English Channel and launched their Folboat a mile from the harbour. Their target was a big German tanker carrying 5,000 tons of copper ore, a vital cargo for the German war effort. Montanaro and Preece paddled cautiously through, and in some places over, the harbour defences and froze when searchlights swept the waters. They attached limpet mines to the German ship and then slid away. It was a dangerous trip back to the MTB because the sea had come up and the wind was against them: as they were hauled on board the MTB their Folboat sank. It was worth the risk. The limpet mines exploded and sank the ship as Montanaro and Preece watched from the MTB while it was en route for Dover. Captain Montanaro won the DSO and Trooper Preece the DCM for their night's work.

The SBS made so many raids that it would be impossible to describe all of them even in a large volume devoted to the unit alone. It is also difficult to nominate a 'typical' raid since every mission was different in its target and its difficulties. What was 'typical' was the audacity, courage and determination of the men. A noteworthy sabotage operation was that carried out in Italy in November 1942 by Captain R.P. Livingstone and Sergeant S. Weatherall. Paddling in after launch by a submarine, they had two objectives. The first was to derail a train in the tunnel at Leignalia in order to block it and therefore prevent movement of supplies and weapons to submarine bases. The second was to report on conditions in the area for possible future operations. Having hidden their canoe in a garden, the pair made their way through the enemy town and reached the railway, where they saw a single sentry on top of the tunnel. Ideally, they should stalk and kill this man but it was more important to keep their rendezvous with their mother submarine. Sub commanders would wait for raiders right to the second of the pre-arranged time, but not a moment later. The captain had no way of knowing if his SBS men had been captured and tortured for information.

With great skill, Livingstone and Weatherall put charges under the rails and on the steel posts carrying the power cables and linked their explosives to a ring circuit. A pressure switch would blow the rails and wreck the train while delayed action charges would create even greater chaos.

On their way out the men found that the country had so many obstructive fences, ditches, glasshouses and buildings that they were forced to use the main road if they were to keep their rendezvous. Boldly, they walked down the road, recovered their Folboat and rejoined the submarine. The captain obligingly stayed on the surface long enough for the saboteurs to witness their explosions. It was a satisfying moment.

Not all raids went so smoothly. Patrols went out to do their business with their lives in their hands. They were on their own, with no way of calling for support or rescue. There may be no better example of this willingness to take risks than the patrol led by Captain D.J. Montgomerie, which was detailed to raid a large German supply dump at Daba, close behind the enemy front line in North Africa. The patrol had to land on an unfriendly coast, move 1½ miles inland and penetrate a large base built around an airfield where fighter squadrons were stationed. The Germans had a sprawling tented camp on the sand hills above the beach and to reach their objective Montgomerie's raiders had to pass through these tents. Similarly, they needed to pass a mobile cinema and a busy canteen. Montgomerie decided to deal with the dump single-handed and in addition he put charges on some trucks captured from the British. Mike Alexander and Peter Gurney put their time-fuse charges on trucks, gasoline barrels and other opportunist targets. They were supposed to rendezvous with Montgomerie on the beach but they did not turn up. Gurney had come across an unseen German post and was wounded by machine-gun fire; Alexander remained with him and tried to get him away but both were taken prisoner. Montgomerie reached safety. The raid could probably be classed as successful but the results hardly seem worth the loss.

In September 1942 Captain David 'Dinky' Sutherland and Captain Ken 'Tramp' Allott carried out an interesting patrol on the enemy-occupied island of Rhodes and while Allott's group was caught this in no way lessens the importance of the venture. It also illustrates the mental and physical toughness of SBS men. Sutherland's target was an airfield used by bombers. From a hiding place on a mountain side he and his 'mucker', Marine Duggan, saw a Savoia-Marchetti aircraft which particularly attracted them as prey. After dark and in driving rain – raiders usually welcomed rain as it was ideal concealment – Sutherland and Duggan reached the airfield but found a sentry standing by the Savoia-Marchetti. The SBS had learned from keen observation that Italian sentries quickly became bored and rarely stayed long in one place and this one was no exception. Even while the man was moving away Sutherland placed bombs on the Savoia-Marchetti and two other bombers. The two raiders wriggled through barbed wire fences, crossed an anti-tank ditch and strolled down a path between buildings. Here they were challenged by a shouting sentry so they slid into the protective darkness. Amazingly, the sentry did not rouse the base. Sutherland and Duggan found a petrol dump, always a desirable target, and laced it with bombs. By the time they had located a hiding place in a river bed the bombs were going off. Simultaneously charges set by Allott and his men began to explode. Sutherland and Duggan counted fifteen separate fires. The aircraft went up in fragments but it was the petrol fire which did most damage because it set light to ammunition which in turn destroyed buildings, vehicles and other stores.

For the Italians, the shock was great. Fire engines rushed to the fires but blazes begun by thermite bombs grow out of control very quickly. Numerous searchlights probed the hills surrounding the airfield and the large garrison fanned out in a search for the raiders. They found Allott and his men. But Sutherland would not run until he saw his damage by daylight. It was widespread and very satisfying. As he and Duggan were moving off they saw a plane land with a general who had come to inspect the sabotage and no doubt to punish the lax sentries. With impressive fieldcraft, Sutherland and Duggan evaded the search parties in country with little cover. During one brief rest they saw an Italian MTB towing the three SBS Carley floats which had brought the patrol to the beach. A submarine was to put in for them but how, now, could they reach it? Their 'safe' area kept shrinking as the hunters closed in. At one time they were crouching under an overhanging rock while fifty enemy soldiers stood nearby and planned a further search.

After three days without food and with water running short, the SBS men were in a precarious position. On the fifth night, weak from hunger and thirst, Sutherland and Duggan were in a spot from where they could begin to send their recognition signals. The wait for the sub's counter-recognition sign was agonising but late that night it came. Sutherland now flashed 'Swimming. Come in.' They swam for nearly three miles. They knew now that they were near death by drowning for they did not have the strength to swim back to shore. They hung on, encouraging each other, and on the verge of collapse they were found by the submarine. They were bundled below and the sub crash-dived. The captain had outstayed the length of time arranged with Sutherland and an Italian destroyer was cruising in the vicinity. It depth charged the sub, shaking it dangerously.

By next morning Sutherland and Duggan were in Beirut where Sutherland, now seriously ill, was taken to hospital. A strong and burly Scot, he survived to lead many more successful raids and in December 1944 he became the highly respected lieutenant-colonel commander of the SBS in the Aegean, Italy and Greece. The nickname 'Dinky'? It had nothing to do with a diminutive stature nor was it an allusion to drawing-room manners. Sutherland was a tough soldier but he was always neat, clean, and polished, in contrast to some of his colleagues who were scruffy and careless about their appearance. David Lodwick, who knew him well, said that Sutherland could emerge from the hold of an island caique after days in dirty and cramped conditions and still look as if he were ready for the parade ground.

The Greek island of Simi, occupied by the Germans in 1941, was one of the unit's most favoured campaign grounds. The raiders even temporarily controlled the island. It is estimated that as many as 18,000 German troops, in addition to the larger number of Italians, were tied down by the actions of the SBS on Simi and other Aegean islands. The very large island of Crete, which the Germans had captured in 1941 following heavy losses in an airborne landing, provided the SBS with many raiding opportunities. On the night of 9 June 1943 a party of three, led by Captain G.I. Duncan, raided Kastelli airfield with spectacular results. The Germans admitted that the SBS raiders' bombs destroyed 7 aircraft, 6 motor transports, 210 drums of petrol and 3 bomb dumps. In addition, they had 70 men killed or wounded. As a bonus for the British, the Germans shot their army guards on duty that night. Unfortunately, they also shot seventeen Greek night-watchmen for failing to spot the raiders, who escaped unharmed.

An SBS raid in Crete on 23 July 1944 was classic in its preparation and execution. Reconnaissance parties provided information about the worthwhile targets, the time needed to get from one target to another was calculated and the best positions for spreading tyre-busters on the roads were chosen. The tyre-busters were an important part of the raiders' gear; judiciously placed they could bring to a halt any pursuit by Germans in cars or on motorbikes. Numerous SBS men took part in the July attack though, as always, in small teams.

Seven separate petrol dumps were destroyed and 165,000 gallons of German petrol went up in sheets of flame. Seven staff cars and numerous trucks were wrecked. Thirty-two Germans were killed and others wounded. SBS losses were an officer and a marine taken prisoner. Despite the careful planning for this raid some SBS men were forced to change targets. Captain Dick Harden and Signaller Stephenson arrived at their target only to find that a large party of Germans were removing the petrol. With too many alert enemy in the vicinity to risk attack, Harden took advice from the local Cretan guerrillas. 'Come and meet Ivan', they said. It is possible that the man's name was Ivan for he was a Russian serving in the German Army who had deserted and was being hidden by the Cretans.

To Harden, Ivan suggested a bargain. He would lead him to a post containing eleven Germans – if Harden would provide him with safe conduct to Egypt. Harden accepted this deal. Ivan guided him to the enemy post where the Germans were at dinner. All were killed in the fusillade fired by Harden and Stephenson. Ivan was eager to do more for Harden. Next morning, he helped the SBS to ambush a German staff car and kill its occupants. With raids such as this, the SBS was successful out of all proportion to its numbers.

The SBS did not confine its activities to raids against German and Italian targets in the Mediterranean. Their patrols also operated against the Japanese in the even more difficult and dangerous country of Burma. There was not even the possibility that they would be taken prisoner should a mission be a failure and sent off to a reasonable life in a POW camp like those maintained by the European enemies. The Japanese would torture them and then behead them. To be operating alone in the jungles and rivers of Burma required men of very fit body and exceptionally strong mind.

On the night of 25/6 November 1944 two Folboats were sent to attack an enemy motor launch that was causing serious trouble along a length of the River Chindwin. One was crewed by Major Sidders and Sergeant Williams, the other by Lieutenant Wesley and Corporal Hickman. Wesley had a Bren light machine-gun mounted in the bows of his Folboat while Sidders and his companion had only their Tommy-guns and grenades. The Japanese craft was known to be armed and the SBS mission seemed like suicide. The SBS men were counting on surprise. After dark the Folboats kept station in the middle of the river and a very short time later they first heard and then saw the enemy boat approaching from the west bank. Unseen because of their black, low profile on the waters, Sidders and Wesley held steady until the boat was no more than 25 yards away. Sidders nodded a command and all four men opened up. This raking fire did indeed catch the Japanese crew by surprise but their captain was able to head for the east bank. Sidders and Wesley had earlier considered their tactics in this eventuality and now Wesley and Hickman cut off the boat's escape in that direction. The Japanese skipper turned back to the west but Sidders and Williams were ready for him.

Out of control, the Japanese boat went around in crazy circles. Its movements were almost too erratic for the SBS men to get alongside but after several tries they succeeded in boarding it. They found that eight of the crew of nine had been killed by the small arms fire; the other man, a lieutenant, was wounded but as he still showed fight he was bound. The SBS patrols had captured a 20-foot boat which was still in good condition despite the British bullets. They lashed their Folboats to the deck and steamed for the British lines. The captured boat was carrying nothing but rice, probably intended for garrisons on the Chindwin's east bank. The Japanese lieutenant proved to be a useful interrogation subject. Sidders and Wesley and their men had been outstandingly successful. They had captured an enemy boat and killed the crew without casualties to themselves. The incident had absolutely no effect on the Burma campaign but it is fair to speculate that the Japanese along the Chindwin, who had already been subject to many harassing raids, became even more alarmed, especially when their British enemies made their small victories look so easy. The SBS made about ninety sorties in total on the west coast of Burma.

Without doubt, the most successful SBS raider, as well as its most flamboyant was Anders Lassen, briefly mentioned elsewhere in this book. His decorations show the status of this young Dane serving in the British Army – the Victoria Cross and three Military Crosses, all earned while serving in the SBS. Lassen, from an aristocratic family, was at sea as an ordinary seaman when the war broke out. After Denmark was overrun by the Germans he volunteered to return as a spy and saboteur. His offer was taken seriously and he attended a special training school in Scotland but there his officers suggested that he would make a better commando than a spy. For spying he was just a little too noticeable, being nearly 6 feet 6 inches in height and rather gangly. Also, he was not self-effacing enough to operate undercover.

Under Captain Gus March-Phillips he was a member of a raiding party in West Africa before he finished up with the SBS under a succession of COs, including George (Lord) Jellicoe, David Sutherland and John Lapraik. A handsome, modest man, Lassen was the ultimate warrior and ideally suited for the independent actions in which the SAS and SBS excelled. The Greek, Aegean and Italian partisans with whom he worked idolised him and he became known as the SBS's 'terrible Viking'. For the German and Italian troops he was indeed terrible: he personally killed more enemy than any British soldier of any rank. Perhaps his hatred for the Germans was sharpened by their invasion of his homeland.

Operating alone and without prospect of help if a raid went wrong, a patrol leader needed to be incisive and dynamic. He had no time for contemplation but had to act boldly and immediately. A raid led by Lassen on the Aegean island of Santorini in April 1944 shows he had these qualities and reveals him at his most ruthless. While other SBS teams were busy elsewhere on the island, Lassen chose Santorini township as his target. Here he found that the Bank of Athens had become the billet for 48 Italian troops and 20 Germans, a number greatly in excess of his 12 men. The raiders had to evade sentries and guard dogs but in the darkness this was no problem for men as well trained and as experienced as Lassen's patrol. The enemy leaders had been guilty of poor judgement in selecting the Bank of Athens for the front door was the only entrance–exit. No SBS man would have so carelessly trapped himself.

The party crept silently into the building and, with Sergeant Nicholson, Lassen padded from room to room. These two had worked together many times and they had a technique for dealing with enemy-occupied buildings. Nicholson kicked open a door, Lassen threw two grenades into it and Nicholson, waiting only the few seconds for the explosion, sprayed the room with his Bren machine-gun. Finally, with his pistol Lassen shot any survivors. Some terrified Italians jumped from windows 40 or 50 feet above the street. It is just possible that they were among the six Italians and four Germans who survived Lassen's massacre. It can be described in no other way but Lassen would have objected to the word: 'These enemy soldiers were all armed and they could have shot me.' Indeed, Lassen's friend, the Greek Stefan Casulli, and Sergeant Kingston, the medical orderly, were shot and killed and two other raiders were wounded. Neither Lassen nor any other raider ever gave an enemy an even break, though if they shouted their surrender fast enough and loudly enough they were usually spared.

Exploring the outskirts of Salonika, Greece, in October 1944, Lassen came across four civilian fire engines in good condition. Ever the opportunist – and exhibitionist – Lassen commandeered them and ordered his men to mount up. Then he led the way into the centre of the town. The Germans were preparing to evacuate Salonika and they took no notice of the fire engines until Lassen gave his raiders the signal to open fire. They inflicted sixty casualties. Lassen himself killed eight of the enemy while

Lieutenant J.C. Henshaw killed eleven. Pleased with themselves, the patrol drove off without casualties to themselves.

The range of Lassen's activities and his consistently outstanding leadership and courage are shown by the citations for his three Military Crosses:

THE MILITARY CROSS

SECOND-LIEUTENANT ANDERS FREDERIK EMIL VICTOR SCHAU LASSEN

Second-Lieutenant Lassen has at all times shown himself to be a very gallant and determined officer who will carry out his job with a complete disregard for his own personal safety. As well as, by his fine example, being an inspiring leader of his men, he is a brilliant seaman possessed of sound judgment and quick decisions. He was coxswain of the landing craft on an operation and effected a landing and subsequent re-embarkation on a dangerous and rocky island with considerable skill and without mishap. He took part in a further operation on which he showed dash and reliability. He recently took part in another highly successful operation in which he was the leader of a boarding party. Regardless of the action going on around him, Second-Lieutenant Lassen did his job quickly and coolly and showed great resource and ingenuity. Second-Lieutenant Lassen also took part in another operation, as bowman on landing, and then made a preliminary reconnaissance FOR A REPORTED MACHINE-GUN POST.

SECOND MC FOR LIEUTENANT LASSEN

This officer was in command of the patrol which attacked Kastelli Pediada aerodrome on the night of 4th July. Together with Gunner Jones, J. (RA) he entered the airfield from the west, passing through formidable perimeter defences. By pretending to be a German officer on round he bluffed his way past three sentries stationed 15 yards apart guarding Stukas. He was, however, compelled to shoot the fourth with his automatic and in so doing raised the alarm. Caught by flares and ground searchlights he was subjected to very heavy machine-gun and rifle fire from close range and forced to withdraw. Half an hour later this officer and another rank again entered the airfield in spite of the fact that all guards had been trebled and the area was being patrolled and swept by searchlights. Great difficulty was experienced in penetrating towards the target, in the process of which a second enemy sentry had to be shot. The enemy then rushed reinforcements from the eastern side of the aerodrome and forming a semi-circle drove the two attackers into the middle of an anti-aircraft battery, where they were fired upon heavily from three sides. This danger was ignored and bombs were placed under a caterpillar tractor which was destroyed. The increasing numbers of enemy in that area finally forced the party to withdraw. It was entirely due to this officer's diversion that planes and petrol were successfully destroyed on the eastern side of the airfield since he drew off all the guards from that area. Throughout this attack and during the very arduous approach march, the keenness, determination and personal disregard of danger of this officer was of the highest order.

THIRD MC FOR LIEUTENANT LASSEN – SECOND BAR TO MC

This officer, most of the time a sick man, displayed outstanding leadership and gallantry throughout the operations by X Det [X Detachment] in Dodecanese, 13 Sep. 43 to 18 Oct. 43. The heavy repulse of the Germans from Simi on 7 Oct. 43 was due in no small measure to his inspiration and leadership on the one hand and the highest personal example on the other. He himself, crippled with a badly burned leg and internal trouble, stalked and killed at least 3 Germans at the closest range. At that time the Italians were wavering and I[1] attribute their recovery as due to the personal example and initiative of this Officer. He continued to harass and destroy German patrols throughout the morning. In the afternoon he himself led the Italian counter-attack which finally drove the Germans back to their caiques with the loss of 16 killed, 35 wounded, and 7 prisoners, as against a loss on our side of one killed and one wounded.

Lassen's last patrol – it was also to be the final SBS operation – took place in April 1945. No. 2 Commando Brigade was to attack German positions at Lake Comacchio, on the eastern flank of the Allied line in Italy. The lake is an expanse of uninviting water and its depth of 2 feet meant that assault boats could not cross; they drew much more than 2 feet when manned and loaded. It was known from local people that the lake had channels which were deep enough and Lassen, in a Folboat, scouted them. He and his minder, Guardsman O'Reilly, were nearly captured. The SBS men accurately charted the channels and on the night of 8/9 April they guided the Commando units and put them into position for their attack. Lassen's own orders, as a preliminary for the assault, were to take his patrol about 2 miles to the north and, in the words of the well-known Army phrase, to 'create a diversion'. Creating diversions was very much to the liking of 'the terrible Viking'. Leading his patrol, he advanced along a road towards Comacchio township and encountered enemy strong points. Already alarmed by firing from the south, the defenders were ready and opened fire. One man was killed and O'Reilly badly wounded. The others found what cover they could but Lassen went on alone against the strong points. One he silenced with a grenade and thus encouraged he continued his advance and was now joined by his men. Lassen had the knack of filling his men with his own confidence. All by himself he put three more strong points out of action with grenades and his Tommy-gun. At the fifth the Germans hung out the white cloth of surrender. Lassen went in closer to accept it and was shot at point-blank range. His enraged men killed every German in the post. They were carrying Lassen back towards the Folboats but he indicated for them to stop. 'I'm done for,' he said. Almost immediately he died from his wounds.

Posthumously, he was awarded the Victoria Cross. His citation is worth quoting in full and the testimony of Sergeant-Major Leslie Stephenson, who was with Lassen on the raid, adds to its graphic nature. Few VC citations are so replete with detail:

MAJOR (TEMPORARY) ANDERS FREDERIK VICTOR LASSEN MC

In Italy, on the night of 8th/9th April 1945, Major Lassen was ordered to take a patrol of one officer and 17 other ranks to raid the north shore of Lake Comacchio. His tasks were to cause as many casualties and as much confusion as possible, to give the impression of a major landing, and to capture prisoners. No previous reconnaissance was possible and the party found itself on a narrow road flanked on both sides by water.

Preceded by his two scouts, Major Lassen led his men along the road towards the town. They were challenged after approximately 500 yards from a position on the side of the road. An attempt to allay suspicion by answering that they were fishermen returning home failed for when moving forward again to overpower the sentry, machine-gun fire started from the position and also from other blockhouses to the rear.

Major Lassen then attacked with grenades and annihilated the first position containing four Germans and two machine-guns. Ignoring the hail of bullets sweeping the road from three enemy positions, an additional one having come into action from 300 yards down the road, he raced forward to engage the second position under covering fire from the remainder of the force.

Throwing in more grenades he silenced the position which was then overrun by his patrol. Two enemy were killed, two captured and two more machine-guns silenced.

By this time his force had suffered casualties [including O'Reilly and Crouch, already referred to] and its fire power considerably reduced. Still under a heavy cone of fire Major Lassen rallied and re-organised his force and brought his fire to bear on the third position. Moving forward he flung in more grenades which produced a cry of 'Kamerad!' He then went forward to within three of four yards of the position to order the enemy outside and take their surrender.

While shouting to them to come out he was hit by a burst of Spandau fire and fell mortally wounded, but even while falling he flung a grenade, wounding some of the occupants and enabling his patrol to dash in and capture the final position. [This was where his angry men killed the treacherous defenders.]

Major Lassen refused to be evacuated as he said he would impede the withdrawal and endanger lives and as ammunition was nearly exhausted the force had to withdraw. By his magnificent leadership and complete disregard for his personal safety, Major Lassen had, in the face of overwhelming superiority, achieved his objects. Three positions were wiped out, accounting for six enemy machine-guns, killing eight and wounding others of the enemy, and two prisoners were taken. The high sense of devotion to duty and the esteem in which he was held by the men he led, added to his own magnificent courage, enabled Major Lassen to carry out all the tasks he had been given with complete success.

Sergeant-Major Stephenson had already given other details in his report on the action:

I heard him shout and ran over to him. I found him lying two yards in front of the pillbox. I lifted him a little by picking him up under his arms and supported him against my knee. He asked who it was and I said, 'Steve.' He said, 'Good – I'm wounded, Steve, I'm going to die. Try to get the others out.' I was the last one to speak to him and I can testify that he died as he had lived – a very brave man.

Stephenson did get the survivors back to their base on an island in the lake arriving there in the early hours of 9 April. That morning, with the fighting over, Italian pro-Allied partisans found Lassen and other dead SBS men and carried them into the town. They were buried near an old stone wall and girls placed flowers on his grave.

The attack did what it was intended to. Lassen's 'diversion' forced the Germans to withdraw men from their main front to stiffen the garrison at Comacchio. When the attack was made, as planned, 24 hours later, the Commandos' casualties were lighter than might have been expected. The German surrender in Italy became effective on 2 May 1945.

To Lassen's many admirers, comrades and friends it had seemed inevitable that sooner or later he would carry out an exploit worthy of the Victoria Cross and that he would probably die earning it. Some of his close friends said that he expected to be killed in action. He was so frequently in the field that the odds were against his survival. Raiders rarely showed grief but the entire SBS mourned Lassen's passing.

1. The officer making the recommendation was Colonel D.J.T. Turnbull, commanding Raiding Forces. It was countersigned by Lieutenant-General Anderson, commanding 3 Corps and General H.M. Wilson, Commander-in-Chief, Middle East Forces.

9

COMBINED OPERATIONS
PILOTAGE PARTIES

WARRIORS ON THE WATER

Throughout the war of 1939 to 1945, whenever there was a need for raiders to undertake a special project demanding special skills, men could readily be found who would risk their lives for freedom. Usually they did not even have to be sought, for when 'the authorities' finally perceived a need they found that adventurous and thoughtful individuals were already doing the job.

As already described in this book, R.J. Courtney demonstrated the need for a lone raider who could fix a mine to an enemy ship. This led to the creation of the Folboat Troop, which led to the SBS. It was Courtney, with Lieutenant-Commander Clogstoun-Willmott, who realised the need for close-quarter reconnaissance of many targets and for navigating pilotage parties to help make raids effective. It can hardly be said that they were pioneers in making raids against ships in harbours but they were among the first to set about this type of raid, or landing on an enemy shore, in a systematic manner.

Clogstoun-Willmott was from a rare breed; he was a GHQ planner but unlike so many other men doing this work he set about studying his task in the field or, in his case, on the water. He was to be the navigator for a raid being prepared against enemy positions on the island of Rhodes. He and Courtney were teamed. They might have seemed like an 'odd couple' – Courtney big, hearty and extrovert, Clogstoun-Willmott quiet, more introspective – but their partnership was fruitful.

Clogstoun-Willmott surveyed Rhodes' beaches through a submarine periscope and although his practised eyes took in a lot he realised that he still did not know enough to plan a raid. He and Courtney, who had already decided that a reconnaissance canoe was better launched from a submarine than from an MTB, paddled ashore from the sub HMS *Triumph*. One of them would hold the canoe close inshore while the other collected samples of the shingle, located submerged rocks, measured gradients and observed any surf. All this required concentration but both men also had to be constantly alert because this was a guarded shore. There was no knowing whether inquisitive civilians would report their activities to the police.

The two officers carried out these tasks for three days. On one occasion they came ashore just in front of the main hotel where Germans were staying. There were some tense moments because the operator who had swum ashore found it difficult, on every occasion, to find the canoe when he swam out. Then they had trouble finding their submarine: a low silhouette was desirable but not to the point of invisibility when working with friends.

The work of Courtney and Clogstoun-Willmott in beach reconnaissance should have been put into effect before the Dieppe raid. But, better late than never, a group with the odd code name of 'Party Inhuman', came into being. Its first recruits came from the SBS and from the Admiralty's hydrographic branch. 'Party Inhuman', with little training, was hurried to Gibraltar, there to work with the planners of Operation Torch, a large-scale landing on the French North-West Africa coast. Unfortunately, but understandably, the naval chief at Gibraltar ordered that no canoe operations could take place; spies abounded in North Africa and should canoeists be observed the Torch landings would be compromised.

This countermanding order was unfortunate because the canoes' crews could have gone a long way to solving the chronic shortage of high-grade navigators. Thousands of Allied craft of all kinds were now in service and many landings were imminent but most naval navigators were poor at their job. They could navigate between two points but close inshore work was beyond them.

Everyone now recognised the need for Combined Operations Pilotage Parties (COPP), as the British called them. The Americans knew these specialists as Amphibious Recon Patrols and Beach Jumpers. From the beginning of the planning for the invasion of Normandy, COPP teams were busy. About 250 men served in COPP and of these at least 50 were lost. This casualty rate indicates the hazards of the work.

COPP, commanded by Clogstoun-Willmott, had an important supporter, Admiral Lord Louis Mountbatten. He, more than most people, knew the difference that COPP teams could have made at Dieppe, and he wrote to the Chiefs-of-Staff to urge that COPP be given priority in terms of equipment and facilities:

The reasons of security have made it impossible to tell all concerned why small demands for individual items are urgently necessary. It is therefore proposed that the CCO [Chief of Combined Operations] should be authorised to use the word 'COPP' as a codeword to obtain the highest priority, and it is requested that the Service ministries may be invited to accord the highest priority on demands prefixed by this codeword.

Consequently, COPP parties were issued with the latest equipment, some of it developed specially for them. They had underwater writing pads, luminous watches and compasses, high-power underwater torches, infrared homing gear, augers for extracting sand and soil samples, night binoculars, tough insulated swimwear and much more. Many of these items did not become publicly available for decades after the war.

Five months were needed to train a team of COPP raiders, a timescale which made these men invaluable. The loss of a team was a major setback. Seven officers, taken by submarine from Gibraltar, were sent to recce Sicilian beaches. Many difficulties cropped up and only two officers, Lieutenant P.R. Smith and K.P. Braund, both naval officers, returned from the mission. These two paddled 75 miles to base, an exertion that only truly fit men could live through.

During the planning for Operation Overlord, the Normandy invasion, COPP made one of the great raids of the war and probably the least publicised. At one point in the preparation the planners became very worried. Air photographs of a key part of the intended landing area showed odd dark strips, which soil experts suggested could indicate peat bogs. Such ground was notoriously unstable and men, tanks, guns and vehicles would stall in it. There was a desperate need to know precisely what those dark patches signified. French Resistance could give no useful information and its men would not risk the danger of being spotted digging holes in this well-defended region. The only men who could provide the answer were the COPP teams.

The problem was handed to Clogstoun-Willmott, who chose to lead a recce raid himself. He took a party of five, including two very strong swimmers, Major Scott-Bowden and Sergeant Ogden Smith. On 31 December 1943 they travelled in two Landing Craft Navigation (LCN), which were towed a certain distance towards the French coast. This still left Clogstoun-Willmott and his men 3 hours from the point where he proposed to put his swimmers in the water. Then the weather turned foul, with turbulent seas and driving spray. The swimmers suffered acute seasickness and they seemed to be much too unfit to make the exhausting swim. Nevertheless they dissuaded Clogstoun-Willmott from aborting the mission and bravely slipped into the icy water. There was a scare when a lighthouse beam came to life but it soon went out and by midnight Scott-Bowden and Ogden Smith were ashore and taking soil samples, on the beach and inland.

Evading numerous patrols and sentries, the COPP men swam back to the LCN with their sample bags and equipment. They were utterly exhausted when pulled aboard but their labours had not been in vain: the geologists and soil specialists could now say positively that there was no peat bog in which Army tanks and trucks could be trapped. Scott-Bowden and Ogden Smith made further reconnaissances to provide more information for the D-Day planners. They and other COPP teams marked the routes for the invasion itself and then became pilots to guide units to the right place on the correct beaches.

The effectiveness of the COPP men was shown in a negative and tragic way. The main American landing was to be on the beach code-named Omaha but the officers in charge of this major operation rejected COPP's offer to place markers in the sea; they said that the Germans might spot them and draw the obvious conclusion that an invasion was imminent. As a result, the Omaha landings turned into a near-disaster in which many men drowned. Scores of the landing craft and amphibious tanks foundered. Also, because of the lack of markers thousands of American soldiers were taken ashore at points up to a mile from where they were supposed to be. And all this after thorough COPP reconnaissances earlier in the year, which the Americans had ignored.

None of the troops taking part in other momentous landings in June ever got to hear about the COPP work that had taken them safely to the shore and beyond. British and Canadian troops landed on Gold, Juno and Sword beaches with few casualties largely because of the efficient pre-invasion work of COPP teams.

In Burma, the COPP teams carried out hundreds of missions, some of them requiring much technical knowledge and skill. For instance, they had to obtain detailed and comprehensive information about tides in the bays and estuaries where the Army and Navy were to operate. There was always some new problem. For instance, an Army patrol reported that they had found 'stakes' driven into the sand just above the low-water line in a bay where British units were to make an assault against Japanese positions. The report was referred to COPP and on the night of 11/12 January 1945 a party of six in three Folboats set out to make a further investigation and if necessary, destroy the stakes. Launched from a light naval craft, the raiders had a long and tiring paddle to reach the area noted by the Army patrol and found that the obstacle was much more formidable than mere stakes. They were poles of tough teak, 15 feet in length and up to 12 inches in diameter. Each had a pointed end and they were spaced in such a way that it would have been impossible for landing rafts and barges to avoid them. The craft would have had their hulls ripped open and the rafts would have capsized.

The COPP Folboatiers attached explosives to the poles and linked them to a ring circuit finishing in a delayed action fuse. In the subsequent explosion twenty-three poles were destroyed, leaving a wide gap for the British Commandos to make their landing. The COPP teams' task was not yet over; they then piloted the Commandos' boats through the gap. Without their services the attack would have failed and there would have been many British casualties.

1 0

TWO TOBRUK RAIDS

TWIN PINEAPPLES, 18/19 JULY 1941. OPERATION
AGREEMENT, SEPTEMBER 1942.

Many raids were made against enemy positions around Tobruk,
Libya, when the Italians and Germans were besieging the town. Later,
raids were made inside Tobruk after the Axis forces had captured it
from British forces under a South African general. Two raids on the
town are worth comparing because the first was a classic of its kind
and totally successful, while the second was such a failure that a
description of it might well be subtitled 'How not to make a raid'. It is
only fair to say that the second raid was infinitely more complex than
the first and this was indeed a major reason for its failure.

TWIN PINEAPPLES

No. 8 Commando was ordered to capture an Italian strong point on
the Tobruk perimeter, called Twin Pineapples because of the shape
of the land where the post was situated. A party of three officers
and forty men, together with a detachment of Australian combat
engineers, was sent to do the job.

Most of the Commandos had come from Guards regiments and
so had strong physiques, with the result that they could march
strongly and carry weights. This does not mean that shorter men
were somehow inferior – they were not. However, on this occasion
it was noticeable that almost every raider was a big man. This
could have a disturbing effect on the enemy in the event of hand-
to-hand fighting. The party's commander, Captain Mike Keely, was
himself from the Devonshire Regiment, his second-in-command,
Captain P. Dunne, was from the Royal Horse Guards and the other
officer, Lieutenant J.S. 'Jock' Lewes came from the Welsh Guards. He
was to become the training officer of the SAS, under David Stirling.

Half the party was armed with rifle and bayonet, the other half
with Tommy-guns. All carried a supply of grenades while one man
in three packed a groundsheet to be used as a stretcher for
casualties. It was considered unwise to take unarmed stretcher-

Five warriors who received VCs from King George VI at Buckingham Palace on 22 June 1945. Standing (from left) are Lieutenant Basil Place, Lieutenant-Commander Stephen Beattie, Lance-Corporal Harry Nicholls and Lieutenant Donald Cameron. Seated is Major Frederick Tilston. Other than Nicholls, all were raiders. (Tilston family)

Admiral of the Fleet Sir Roger Keyes, first Chief of Combined Operations, who encouraged Commandos and other raiders. (Admiralty)

Lieutenant-Colonel Geoffrey Keyes, who was killed while leading a Commando raid on a building where General Erwin Rommel, 'the Desert Fox', was supposed to be. The VC was awarded to Keyes posthumously. (Keyes family)

David Stirling, probably the greatest raider leader of the war. This is an enlargement from a group photograph. (SAS)

Lieutenant-Colonel Blair 'Paddy' Mayne, David Stirling's successor as leader of the SAS and four times awarded the Distinguished Service Order.

The SAS leaders depended on the Long Range Desert Group navigators to lead their patrols out and safely bring them in. Here two patrols meet in the trackless desert. This photograph was taken by an LRDG soldier.

Men of the Long Range Desert Group on patrol in North Africa. They covered vast distances in their raids against the Germans and Italians.
(David Lloyd Owen)

An advanced dressing station and a wounded Commando during a Norway raid. (Ministry of Information)

A scene during the Lofoten raid. The soldier on the right is Major (later Brigadier) Peter Young, one of the great Commando leaders of the war. (Ministry of Information)

Commandos leap from their Landing Ship Infantry to attack German positions at Vaagso. Notice the men with scaling ladders. (Ministry of Information)

Commandos admire their handiwork – a great oil fire during the Vaagso raid. (Ministry of Information)

An MTB. Many raids were planned around the use of Motor Torpedo Boats which were fast, dependable and heavily armed. They landed and retrieved raiding parties. (Navy)

Anders Lassen (standing), the 'Terrible Viking' to the Germans and Italians in the Aegean Sea, Italy, Albania and Yugoslavia. He was the most decorated raider – VC and three MCs. (Lassen family)

The Würzburg radar dish (left) and the German radar base at Cap Bruneval. Commandos stole vital parts of the installation so that British scientists could study them. (RAF Photographic Reconnaissance Unit)

The radar dish after the raid.

American raiders pinned down in coastal swamps by the Japanese defenders of a Pacific island. Raiders could not afford to stay down for long. (US Marine Corps)

The classic Folboat canoe with its two raiders using double-bladed paddles. Apart from the men, a Folboat could carry up to 400 lb in weight, mostly taken up with demolition charges. (Imperial War Museum)

Commandos make an assault on Dieppe (see Chapter Seventeen).
(Imperial War Museum)

Survivors disembarking after the Dieppe raid. (Imperial War Museum)

Commander Robert Ryder, who led the Commando–Naval raid on St Nazaire. He was one of five VC winners. (Admiralty)

Lieutenant-Commander Stephen Beattie who rammed his destroyer, HMS *Campbeltown*, into the lock gates at St Nazaire. He was captured but awarded the VC. (Beattie family)

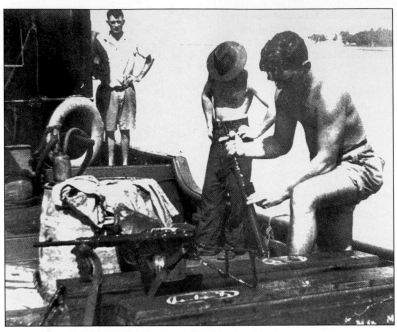

Members of the Jaywick team check their weapons on the *Krait* shortly before its departure for the Singapore raid. (Australian War Memorial)

The pilot of a 'Sleeping Beauty', as used by Lyon, submerges his craft by letting water into two buoyancy tanks near his legs and trims using a small tank in the bows. (Australian Archives)

Captain Jock McLaren raided
Japanese positions by land and
sea. The enemy put a price on
his head. (Rex Blow)

One of the greatest raiders of
the war – Major Rex Blow. He
risked his life every day for
three years. (Rex Blow)

The two-man Chariot which was modelled on the Italian Maiale, usually referred to as 'the pig'. (Imperial War Museum)

Mark I human torpedo is hoisted aboard its mother ship. No warhead had been fitted in this photograph. The craft's bows are at the top of the picture. (Imperial War Museum)

The Storch aircraft in which Otto Skorzeny flew Mussolini from his mountain prison. (Published in the German magazine *Signal*)

Adolf Hitler greets his friend and ally Benito Mussolini after his rescue. (Published in the German magazine *Signal*)

Group Captain Charles Pickard (left) and Flight Lieutenant Bill Broadley prepare for departure to lead the Amiens prison raid from which they failed to return.

Amiens prison after the bombing. (Squadron Leader Tony Wickham, RAF Film Production Unit)

bearers on a raid such as this. By now all Commando soldiers wore rubber boots; the hobnailed Army boots were far too noisy.

The raiders set off from their own base near midnight and passed silently through the Italian forward posts and the main defensive positions. There had been no challenge as they reached the track used by the Italians to take rations forward. They were soon in the rear of the Twin Pineapples position. Here they paused, by arrangement, while Indian troops created a diversion by firing flares and shooting at the Italians. The Commandos drew closer and closer to the Italians' rear. From 30 yards out they charged, firing from the hip and shouting 'Jock' – the password. Open fighting lasted about 4 minutes, then the Italians retreated to their dugouts, where the Commandos bombed them.

Mike Keely rushed a machine-gun post and was seen by Sergeant Dickason to 'clout the crew with the butt end of his Tommy-gun; his Tommy-gun was useless after that but so were the enemy gunners'.

The Australian sappers quickly blew up the enemy ammunition dump and within 15 minutes the raiders withdrew, just before Italian artillery opened up on Twin Pineapples. No. 8 Commando had one man mortally wounded and four other casualties, a small loss for such an outright success. That it was such a success was due to careful planning and reconnaissance, decisive leadership and spirited action by all ranks.

Afterwards thirteen half-pint tins of beer were issued to the Commandos. This was more important than it might appear, for every man in besieged and heat-struck Tobruk was forever thirsty. With not enough beer to go round, the three Commando officers surrendered their claim to a share. Since the total amount did not divide equally, lots were drawn for the amount left over after an even issue. The Commando chronicler observed: 'For a brief time this man was the only member of the Tobruk garrison not thirsty!'

OPERATION AGREEMENT

The SAS and other raiding units did not always have it their own way. Agreement illustrates the desperate dangers faced by raiders bold enough to enter the lions' den, which Tobruk had become after its capture by Rommel's forces.

The raid had one objective: to hold Tobruk for 12 hours to allow demolition squads to destroy the large fuel dumps while the Navy sank ships in the busy enemy harbour. In retrospect, it seems more like a raid mounted for its own sake.

LRDG, SAS and Special Interrogation Group (SIG) were all involved but the strike force was supplied by D Squadron, SAS. It consisted of five officers and thirty-two other ranks, all hand-picked for the operation. An LRDG patrol under Captain Lloyd Owen would guide the raiders to Tobruk and then cover their rear. Also, Lloyd Owen was to seize part of a radio direction finder.

Getting the raiders into Tobruk was the task of the SIG, under Captain L. Buck. SIG was an interesting unit, being composed of Palestinian Jews of German extraction. These men, all German-speaking, were to don enemy uniforms and act as an escort for supposed prisoners, actually the British raiders. Safely inside the defences, they would establish a bridgehead, allowing yet other troops to be landed from MTBs. At another point a destroyer would put Marines ashore. At yet another point still more soldiers would land from eighteen MTBs; other forces would come ashore from the destroyers HMS *Sikh* and HMS *Zulu*. David Stirling, Paddy Mayne and other experienced SAS officers would never have wanted to be part of an operation in which so much could go wrong.

In overall command was Lieutenant-Colonel John Haselden, a daring man with a background that gave him many advantages. Born and bred in Egypt, he was a fluent Arabic speaker. As a cotton broker he was much travelled and he knew many Arab tribes as friends. On lone intelligence missions he often posed as an Arab. This was the officer who had carried out the initial reconnaissance for the Rommelhaus raid (see Chapter 11). He was too valuable an officer to risk in a hit-and-run raid against heavily defended Tobruk. He was killed during the raid, and his loss left a large gap in the working of British Intelligence in North Africa.

With so many elements whose officers had not received the benefit of a combined briefing or the opportunity to ask questions it was a certainty that difficulties would arise. However, the British 'prisoners' and their 'German' escort had rehearsed their parts thoroughly; the escort group led by Captain Buck even carried 'correct' forged documents. On 13 September the LRDG patrol led the raiders to Tobruk. At a pre-arranged separation point, the 'Tobruk Commando', as it had come to be called, climbed into its three trucks, with Afrika Korps markings, and headed for the perimeter. Bluff and the escort party's fluent German saw them safely through several roadblocks. One sergeant said to Captain Buck, 'What a pity you don't have many more British prisoners, but this lot is better than nothing.'

At Mersa umm Sciause, the raiders debussed as planned to form a bridgehead. They split into two parties, one of which was led by

Lieutenant-Colonel Haselden. The action now began and the Commandos succeeded in establishing their bridgehead, but while most of the Italians they encountered surrendered the Germans as always, fought back.

The Navy's part in Operation Agreement was a fiasco. Navigation was grossly at fault and only two MTBs came into the bay to unload their troops. The boats were late in keeping their rendezvous and when the Germans switched on scores of searchlights and illuminated the bay as if it were a stage there was no further chance of MTBs getting in without being shot to pieces. The destroyers *Sikh* and *Zulu* transferred their Marines to lighters they had towed but the sea was high – which the naval officers must have known – and the lighters broke up, many Marines being drowned. In the dark the captain of the *Sikh* had not seen the fate of the unfortunate Marines and took his ship dangerously close inshore so that on the return journey the lighters would not have so far to travel. Shore batteries pounded the now very visible ship and it caught fire. Many Marines were burnt to death. HMS *Zulu* came to *Sikh*'s rescue and bravely took it in tow. With HMS *Coventry* as escort, both destroyers made for Alexandria. Not one of the ships made it safely through 'Bomb Alley', the Tobruk–Alexandria run. German aircraft sank them with heavy loss of life. When the destroyers departed the Navy knew that Operation Agreement had failed, but communications were so lamentably poor that it was some time before the defeat was obvious to Colonel Haselden. He saw that some of the MTBs were gamely trying to reach the bridgehead under heavy fire and until their commanders desisted Haselden had to keep the bridgehead open.

As dawn approached and the whole ghastly shambles became evident, Haselden gave 'Tobruk Commando' the order to shoot its way out. The men captured some enemy artillery pieces and these they now spiked. The SIG men – the make-believe Germans – destroyed the fake Afrika Korps trucks and exchanged their uniforms for others taken off their dead British comrades.

The fighting became desperate: groups were separated and valorous deeds were performed. They could never be recognised with awards because witnesses became casualties themselves. Colonel Haselden led a counter-attack and was shot dead.

Daring and determined escapes were made by small parties. Lieutenant Tommy Langton with four men put to sea in an infantry assault boat but enemy fire forced them back. Langton found his way across a minefield, broke through the perimeter wire and headed for the British lines, 300 miles to the east. Two months later he was picked up by a British armoured car on patrol. His four companions had perished. Lieutenant David Lanark also turned up, leading three men. Ten Marines, survivors of the botched landing, reached safety after two months' privation. Of the 382 Marines who had set off on the operation only 90 survived. Tobruk's fuel dumps had not been destroyed and nothing had been achieved. Apart from the three destroyers, six of the MTBs were sunk in the hazardous coastal waters. The Navy had suffered grievously. Various SAS and SIG men, trained in evasion and survival, got away through the perimeter southwards and rejoined their units.

Prodigious valour could not conceal the obvious fact that Operation Agreement was a total failure. Stirling and the other raider chiefs would have deplored the human loss and would have said, if only to each other, 'Who planned that bloody operation? What did they expect to happen?'

Agreement really was a lesson in how not to mount a raid. These were the major faults:

- No briefing with all officers concerned had taken place.
- Coordination between Army, Navy and Marines was woeful.
- The raid's commander, Haselden, was not in touch with the ships' captains and the marine officers.
- Haselden should never have been there in the first place; he would have been a prize capture for the Germans, who knew his reputation as an intelligence chief.
- The fuel dumps were more properly a target for a small party of SAS infiltrators and SIG German and Italian speakers, or of RAF bombers.
- Naval reconnaissance for the operation was pitifully inadequate. Officers should have known the sea conditions. Some did know, in which case they should have told the Army that it could not count on help from the sea.
- There was no prospect whatever that such a small party could hold Tobruk for 12 hours, the period demanded by the officers who called for the raid.
- An operation of such difficulty and magnitude called for a highly experienced Commando leader. The brave Haselden was out of his depth.

In war histories Operation Agreement has had little mention. This in itself shows that those who arranged it were ashamed of its management. The officers and men were not to blame; they had been asked to do an impossible job.

11

THE ROMMEL RAID, CYRENAICA, 17/18 NOVEMBER 1941

Military planners have always believed that if a significant enemy leader could be killed or captured, an entire campaign might be won against the side that had been careless enough to lose its chief. Supreme leaders, however, are well protected, not only by their own bodyguard but by the many miles of patrolled and occupied territory between them and potential raiders. The most difficult problem of all facing those who might plan a snatch of a senior general is to know precisely where he is at any given time. This calls for Intelligence information of the highest order.

The British had a special interest in neutralising General Erwin Rommel. A brilliant tactician, Rommel was the most formidable leader of any nationality in North Africa during 1940–2. The British and Empire troops knew him as 'the Desert Fox' and they had as much respect for him as did his own Afrika Korps. He proved himself to be better than any British general until Montgomery defeated him at El Alamein in November 1942. But in 1941 British soldiers had no faith in their own generals and were in fact caustically critical of them. Much of this criticism was unfair and uninformed but rank-and-file soldiers think in black and white terms and at that time in their minds, Rommel was unbeatable.

Colonel Bob Laycock, the most senior Commando officer in North Africa, believed that he could turn the course of the war by removing Rommel, one way or another. The Afrika Korps would be headless and Rommel's 'invincibility' would be shown to be an illusion. Reconnaissance intelligence reports – and Laycock controlled the units which supplied this intelligence – placed Rommel in one of his headquarters. This HQ, 'the Rommelhaus' as it was called, was in Beda Vittoria above the coastal escarpment of Cyrenaica, Libya. An operation was planned to 'take out Rommel'. In overall command was Laycock himself, while Lieutenant-Colonel

Geoffrey Keyes, of No. 11 Scottish Commando, volunteered to lead the assault on the Rommelhaus. Son of Admiral of the Fleet Sir Roger Keyes, Chief of Combined Operations, Geoffrey was following his father as a dynamic leader.

Eighth Army HQ was unequivocal in its directive to Laycock and Keyes: Rommel was to be captured and brought out or he was to be killed. Since there could be only one chance in a million of getting him back to British lines, the directive was tantamount to a death sentence. To give such a positive order was a dangerous step because it invited retaliation in kind.

The raid was to take place on the night of 17/18 November and was timed to coincide with the opening of General Auchinleck's campaign to relieve besieged Tobruk. The operation was hazardous and Laycock knew it. He paraded his party of six officers and fifty-three men and said, 'This attack means almost certain death for the assault party and the chances of any of us being evacuated after the operation are only a fraction higher than nil.' Despite the grim warning, the Commandos remained cheerful and optimistic and Keyes privately urged Laycock, his chief, not to repeat his forebodings in case generals in Cairo became worried and cancelled the operation.

In the submarines *Talisman* and *Torbay*, the Commandos reached their rendezvous off the Libyan coast and almost at once saw the torch-flash that indicated the coast was clear for them to land. Holding that torch was a renowned raider Intelligence Officer, Captain J.E. Haselden of the Desert Reconnaissance Group, who had previously dropped by parachute behind the enemy lines. He spoke fluent Arabic and in disguise often passed as an Arab.

The landing was a near disaster because of rough water. Some of the rubber boats were swept away as were a number of men. Only half the party reached shore, a terrible beginning to the mission. Hiding his party in a wadi to spend the daylight hours, Laycock divided the men into three groups. He, with a sergeant and two men, would stay with the dump of stores and ammunition. Lieutenant Ray Cook and six men were to cut telephone wires to prevent the alarm from being raised and to isolate the Rommelhaus. Keyes and Captain Robin Campbell led the assault group.

During the next two days the whole party, less Laycock and his small group, slept by day and moved by night. With information eagerly provided by an Arab boy, Keyes made a reconnaissance and produced a detailed map of the Rommelhaus and the area around it. The weather was foul but Keyes considered this to be good cover for his final approach, which involved scaling the precipitous Libyan escarpment.

When the raiders reached the track that was said to lead direct to Rommel's back door, Keyes ordered Cook to proceed separately on his mission of telephone line cutting. Keyes and Sergeant Terry scouted ahead towards the Rommelhaus while Campbell brought up the rest of the party. The night was dark, damp and cold and with Tommy-guns held forward and safety catches off, the Commandos approached to within 100 yards of the quiet building.

It was 11.30 p.m. A dog barked and an Italian soldier emerged from a hut with an Arab. Captain Campbell, a German speaker, told the Italian that he was in command of a German patrol. Corporal Drori, fluent in Italian, repeated this statement and the Italian soldier and his companion turned back into the hut. Keyes signalled his group to take their pre-arranged places for the attack. Followed closely by Campbell and Terry, he broke through a hedge into a garden, rounded a corner and found the first door open. Running up the stairs behind it he came to glass-topped doors, through which emerged a German officer wearing an overcoat and steel helmet. Stepping straight into Keyes's path, this man bravely seized the muzzle of Keyes's Tommy-gun and the two men wrestled for it. Before Campbell or Terry could get round behind him he retreated until his back was against a wall. As Keyes could not draw his fighting knife, Campbell, alarmed by the noise they were making, shot the German with his revolver. Surprise was now quite impossible and Keyes at once ordered his men to use Tommy-guns and grenades.

Enemy soldiers could be heard moving. Keyes opened a door on to an empty room and pointed towards light shining from under the next door. As he crashed it open, he, Campbell and Terry saw about ten Germans in steel helmets. For some reason Keyes now fired a few rounds from his Colt .45 revolver, rather than a burst from his Tommy-gun. Campbell said urgently, 'I'll throw in a grenade!'

Keyes first slammed the door shut and then reopened it to allow Campbell to toss in the grenade. 'Well done!' said Keyes, but they were his last words. A German fired a bullet which hit Keyes near the heart. He died while Campbell and Terry were carrying him outside. In the confusion a Commando mistook Campbell for a German and shot him, badly wounding him in the leg.

Grouping at the rear of the Rommelhaus, the raiders said that they would carry their disabled officer to the evacuation beach. As this was 25 miles away Campbell knew that the suggestion was out of the question and he ordered the men to leave him there. Sergeant Terry now commanded and led his party at the run to what he hoped was safety.

The Germans soon found Campbell, treated him well and took

him to a field hospital where his leg had to be amputated before he was sent to a POW camp. The Germans gave Keyes a funeral with military honours; a senior German chaplain conducted the service. The competent Sergeant Terry and his men reached Colonel Laycock in the base wadi, but Lieutenant Cook failed to appear. He had carried out his task but was captured on the way back.

The submarine *Torbay* returned to meet its rendezvous but the sea was still wild and Laycock and the others could not reach it. Hiding in caves, Laycock and his Commandos saw Italian and German troops in large numbers searching for them in all directions. Short of making a bold but suicidal charge, Laycock had no option but to order his men to break into small groups, sprint across the open ground and find refuge in the hills. They could try, if they wished, to signal the sub *Talisman* or they could lie low until the British overran the area in the impending offensive. Neither choice offered much chance of survival.

Laycock and Terry, staying together, ran the gauntlet of marauding, murderous Arabs and patrolling enemy, and set a course to reach the Eighth Army. Living on berries for days at a time to eke out the few rations they carried, they marched for 41 days before reaching British posts at Cyrene. Since rain was continual they were never thirsty but they were starving when they stumbled into safety. Theirs had been one of the great Commando feats of endurance and they were the only members of the raiding party to get back, other than those taken prisoner and released after the war.

Rommel had never been in danger. The house which the raiders attacked was the HQ of the German and Italian supply services. Rommel may have visited it but at no time did he occupy it. At the time of the raid Rommel was at Gazala, in the forward area, with his soldiers, as he so often was.

Twenty-four at the time of his death, Geoffrey Keyes was awarded the Victoria Cross and lies buried in Sollum War Cemetery with other British soldiers. The raid had been a daring exploit but from the beginning its planning was vitally flawed by errors in Intelligence. The Intelligence Service should at least have known that Rommel was not a general to spend long at his base and if he had heard from his own spies that a British offensive was likely – and he would have heard – he would at once have gone to the front.

Forever after the Rommel raid, planners of other special operations insisted on verification of intelligence reports. I believe that Laycock, Keyes and the staff of Eighth Army HQ were so beguiled by the prospect of doing away with the man who was causing them so much difficulty and humiliation in North Africa that all were ready to believe, on flimsy evidence, that the house at Beda Vittoria was his HQ. Having made his reconnaissance, Keyes must have noticed the almost total lack of security at Beda Vittoria; not a single sentry stood near the back door, which was unlocked. Also, no despatch riders, staff cars or anti-aircraft guns were to be seen. Had the Rommelhaus really been Rommel's HQ it would have been a very busy place. The ease with which he was able to approach the building must surely have puzzled Keyes.

Keyes is said to have written to his family just before setting off on his raid, saying that he did not expect to survive it. He could not have confessed this to his superiors and withdrawn from the mission. He came from a family in which military honour was paramount. Young for a lieutenant-colonel in the Second World War, Geoffrey Keyes was a courageous officer; it was a pity that he had to die to prove it.

12

THE GREAT RAID ON
ST NAZAIRE, 27 MARCH 1942

OPERATION CHARIOT

The British government and the chiefs of the armed forces were justifiably obsessed with the threat posed by the German-occupied French port of St Nazaire on the Loire River. St Nazaire attracted Churchill and his military leaders as a flame attracts moths. They knew that a raid would be difficult, dangerous and costly but they believed that the advantages, should a raid be successful, greatly outweighed the risks. St Nazaire possessed the largest dry dock in the world and the only one that could accommodate the fearsome German battleships *Tirpitz*, *Scharnhorst* and *Gneisenau*. Furthermore it was the most important U-boat base on the French Atlantic coast. Wolf packs set out from here to harry the convoys heading for beleaguered Britain.

Hitler and his High Command well understood the British temptation to attack St Nazaire and strongly protected it. More than 6,000 soldiers and sailors manned 28 heavy guns of the German 200 Naval Atlantic Battalion and batteries of 240 mm railway guns. The Loire estuary bristled with heavy machine-guns and anti-aircraft guns, while patrolling the water of the wide estuary were numerous destroyers, torpedo boats and patrol boats. The dock itself was protected against aerial bombs and naval gunfire by a steel caisson 36 feet thick. The defenders doubted that a naval assault would be made because their alarm systems along the river would quickly pick up any British approach. Anyway, the Germans reasoned, the Mindin sandbanks at the mouth of the Loire would deter the Royal Navy.

Hitler, his generals and admirals had not taken into account British determination and raiding skills. This was surprising because Hitler had already decreed that Commandos were to be shot, such was his fear of these men. The planners intended to smash the caisson with an old former American destroyer, HMS

Campbeltown, loaded with 4½ tons of high explosive, which was well hidden and protected by concrete walls. Even when *Campbeltown* rammed the caisson at full speed the charges would not go off until time pencils detonated them. Accompanying *Campbeltown* were the destroyers HMS *Atherstone* and HMS *Tynedale*, a motor gunboat, a motor torpedo boat and sixteen Fairmile naval launches. The ships carried a raiding party of Commandos under Lieutenant-Colonel A.C. Newman of No. 2 Commando. Included in his force were demolition parties from Nos 1, 3, 5, 9 and 12 Commandos. The naval force was commanded by Commander R.E.D. Ryder and the *Campbeltown*'s skipper was Lieutenant-Commander S.M. Beattie. In all, 611 officers and men set forth on Operation Chariot, including a skeleton crew of 75 on *Campbeltown*.

After the planning, rehearsals and preliminaries, the raid-proper began at 10 p.m. on 27 March, 70 miles south-west of St Nazaire. Here, Ryder, Newman and their small staffs transferred from *Atherstone* to MGB 314, on which they would lead the raiding flotilla into the Loire. The destroyers flanked the small fleet in order to make certain of locating the submarine HMS *Sturgeon*, which had taken station as a marker with a direction light. *Sturgeon* also indicated the destroyers' departure point from the convoy. They were to cover its rear and pick up survivors of the raid. The sailors wondered if there would be any survivors. Early in the operation *Atherstone* and *Tynedale* had attacked a German submarine, which reported the position of the destroyers – unknown to the British officers – but the report did not cause the St Nazaire enemy commandos to suspect a raid.

After a total of 33 hours' navigation the British ships arrived precisely at their take-off point for the attack. From this moment *Campbeltown* guided the raid but, unknown to everybody except the ship's officers, there was the Mindin sandbank which *Campbeltown* must cross. As they waited tensely, the destroyer twice grounded but then slid through the underwater sand. From about this point every man on the operation was alert with expectation of enemy attack. It seemed inconceivable that the alarm had not already been raised and that twenty ships approaching the inner estuary had not been spotted. Still there was no alarm, although some anti-aircraft guns opened up on British aircraft which had been ordered to create a diversion during the raiders' approach.

A searchlight on the bank pierced the sky and slowly descended but its beam fell astern of the convoy before going out. A minute later a dozen searchlights on either bank glared and the water was lit up as if in bright sunlight. Even then the British ship and boats might not have been seen – but the salty foam of their wakes sparkled in the light. *Campbeltown* was now 10 minutes from ramming point. Commanders of each raiding craft quietly ordered 'Full speed ahead'. There was still a chance of covering more distance before the Germans began to shoot. *Campbeltown* had been cunningly altered to make the ship look like a German destroyer and it was just possible that it would be regarded as friendly. It and the smaller craft were flying the German flag, which had been stained and blackened to make recognition difficult from the shore. Before they started to shoot, the skippers were obliged by international law to strike the enemy flag and hoist the White Ensign, the battle flag.

Two signal stations on shore transmitted recognition signals, calling for a response from *Campbeltown*. Leading Yeoman Pike on the motor gunboat replied with illegible flashes, then to one enemy signaller he cleverly gave the recognition signals flashed at him by the other. *Campbeltown* and its attendants ploughed on at speed. When a battery fired some shots over the destroyer's bow, an internationally recognised order to heave-to, Pike put into effect the plan thought out by Ryder and Beattie. In German, he signalled 'Wait', and then gave the genuine recognition signals of a German destroyer, supplied by Royal Navy Intelligence. After a careful pause, to gain more time, he signalled 'Urgent. Two vessels damaged in the course of an engagement with the enemy. Request permission to enter port immediately.'

Another deliberate pause, then Pike signalled 'I will have something else to transmit.' *Campbeltown* was 7 minutes from ramming. Some light batteries opened fire, menacingly this time but not destructively. Now Pike flashed, 'I am a friend. You are mistaken.' The firing ceased.

Six minutes to ramming. *Campbeltown* already had an advantage – at this point the heavy batteries could not open fire so near the port or they would cause great destruction. It would have been better for the Germans had they taken that risk.

The beautiful British bluff could not last forever and soon the defending guns opened up in earnest. Passing an armed trawler that seemed about to intervene at the southern entrance, the MGB fired at it and the following British light craft also attacked it. The German batteries, mistaking their own trawler for an enemy, finished it off with gunfire. By now the German flags had been struck and replaced by the Navy's battle ensign.

Having the trawler destroyed for them was a little luck for the British but there was to be no more luck. The raiding vessels were completely exposed in the searchlights' glare and all they could do was to keep up their own volume of fire, inaccurate though it was. Behind the armour plating on *Campbeltown*'s bridge, Beattie calmly gave directions to the helmsman, making various slight changes in direction so that his destroyer would strike exactly according to plan. *Campbeltown* was heavily under fire now, the Germans belatedly realising its commander's intentions. It was too late. The Navy had calculated that the ship would ram at 1.30 a.m. It was only four minutes late. At a speed of 20 knots and with a great smashing shock – though less noisy than anticipated – *Campbeltown* penetrated 33 feet into the caisson and climbed onto it. It could not possibly be towed off. The damage was immense.

The eighty Commandos on *Campbeltown* had already suffered casualties but this was the moment for which they had been waiting. Unrolling their rope ladders they clambered down the sides of the destroyer onto the dock. Aboard, Chief Engine-room Artificer Howard and Engine-room Artificer Reay opened the valves that would flood *Campbeltown*. The whole dock area was a mass of acrid smoke, flames and fire-flashes and explosions. The din was deafening. In places German guns were firing on the raiders at a range of only a few yards. Other Commandos landed from the launches so that in quick time seven different parties were on their way to the twenty-four objectives of the raid. While this was happening the time-pencil fuses which would blow the tons of explosives hidden within *Campbeltown*'s mangled hull, were silently ticking away.

At the southern end of the dock were three targets of great importance allotted to a small group from No. 5 Commando. Lieutenant Stuart Chant had to destroy the massive pump house; Lieutenant Christopher Smalley would blow up the machinery for moving the great southern caisson in or out of position; Lieutenant Robert Burtenshaw had the task of wrecking the caisson itself. Chant had one of the most difficult tasks of Operation Chariot and he had with him four well-trained sergeants, laden with explosives. One of them, Sergeant Chamberlain, was wounded and disabled but his mates not only carried out their jobs but rescued Chamberlain as well, no easy action on narrow metal stairways. Chant was wounded in the legs but he forced himself to climb down 40 feet, to fix explosive charges to the pumps and to climb back up. Exactly 150 lb of explosives went off with a shattering, earth-shaking roar.

The MGB put Colonel Newman and his small staff ashore at the Old Port's north quay. Commander Ryder was also on this boat but, blinded by smoke and flames, he had no clear idea of what was happening. He met some of the *Campbeltown*'s crew but none could tell him accurately where the destroyer lay. Ryder, with Yeoman Pike as bodyguard, set out to find *Campbeltown* and as they came around a wall, where they sheltered from bullets, they found the ship, with its stern settling nicely after the scuttling process. About this time MTB 74 fired its torpedo-mines against the outer gate of the Vieille Entrée Lock. As had been planned they hit its middle and sank, later to explode.

Ryder was back aboard the MGB when MTB 74 came alongside and took off nine survivors from *Campbeltown*. Ryder ordered, 'Return to England at full speed.' This MTB had a top speed of 40 knots and was the most likely boat to escape from the St Nazaire inferno. Ryder then found that the Germans were still holding the Vieux Mole, around which were battered and burning British MLs. Fire was coming from numerous machine-guns set in blockhouses and from other guns which the Germans had rushed onto nearby roofs. Lieutenant D.M. Curtis, skipper of the MGB, ordered the pom-pom gunner, Able-Seaman W.A. Savage, to take on the machine-guns on the roofs. Under really heavy fire, Savage knocked out some of them and then silenced the guns firing from a blockhouse. Hit by an enemy shell, he died at his gun.

Ryder was now in a difficult position. All around him his force was being destroyed at the hands of an enemy growing stronger by the minute. Ryder had fifty men on the MGB, many of them wounded. If he remained where he was the MGB was doomed and, in any case, there was nothing for him to do. Having consulted Curtis, Ryder gave the order to withdraw. At 24 knots the MGB had a chance of reaching safety but searchlights kept it in focus and the shore gunners continued to spray it with bullets.

The surviving Commandos who had completed their jobs, and the crews of the boats were supposed to assemble at the Vieux Mole for rescue by the MLs but when they reached the Mole they were shocked to see that not a single launch was alongside waiting for them. All over the channels to safety lay shattered hulks, all burning, and all around them and covering the basin was blazing petrol. There was no hope of embarkation. Major W.O. Copland organised a form of defence around some railway trucks, a movement which drew fire from the enemy rooftop gunners. A group of German troops then charged Newman's defences, screaming 'Heil Hitler!' Newman shouted to his men to hold their fire, then with the racing Germans a mere 10 yards away he gave orders to shoot. The Tommy-gunners mowed down the Germans to a man but this gave the Commandos only a brief respite.

Newman and Copland now divided their men into three groups each of about twenty, under Newman, Copland and Captain S.A. Day respectively, with orders to make their way independently through the town. But first most had to run across the great iron turntable of the railway track, while a second party crossed the swivel bridge. Bullets ricocheted with a whine from the steel plates but most of the men survived the dash. Realising that the groups were too large Newman then broke them into twos and threes. However, a large number insisted on staying with him. They dodged from street to street, racing into dark alleys to avoid the German trucks which were hunting down the raiders. Newman found a cellar and here the officers divided up the remaining ammunition and food, though there was not much of either. More importantly, the wounded were bandaged. Then came the clatter of metal-studded boots surrounding the house. A German shouted, 'Come out with your hands up or we'll bomb you out!' Newman had no option but to surrender; to fight would have meant the death of every man in his party.

Meanwhile Ryder's MGB was heading up the estuary for the sea. The Germans, still firing at a smoke float in the belief that behind it lay the enemy craft, did not probe the estuary with searchlights. The MGB caught up with a much slower ML and Ryder, unable to help in any other way, ordered a smokescreen for the smaller boat. Finally the batteries did find the MGB and straddled it with shells, but Lieutenant Curtis at the wheel zigzagged it clear. He faced yet another hazard – a German E-boat, fast and heavily armed, came at the MGB. The British boat now had only one operable gun, an Oerlikon in the bows and as Curtis steered towards the E-boat this gun came into action, setting the German ship ablaze.

As dawn broke, Ryder's MGB was an appalling sight. The bridge was covered with wounded men, huddled against anything that gave them support, and the deck was running with blood. The hull and superstructure had been pierced by small shells and splintered by shards and the vessel was a mass of wounds of its own. HMS *Atherstone* was a welcome sight but the crew had difficulty in hoisting aboard the badly wounded men without causing further pain and without aggravating their wounds. HMS *Tynedale* took aboard survivors of the MLs. Four other MLs reached the open sea but only four of the eighteen that had set out on Operation Chariot finally returned home.

Lieutenant-Commander Beattie was among the 200 survivors taken prisoner and many of these were wounded. Some prisoners were shot on the spot. Private McCormack, fatally wounded in the head, was a piteous spectacle as he lay in an open space in St Nazaire town, with his bloodied head between his kilted knees. He was obviously dying and in need of comfort but the Germans, military and civilians, gathered around him in crowds, jeering and laughing. Their cameras clicked and one shot of McCormack was published throughout Europe in German armed forces magazines and the multi-language *Signal*, with the title 'Picture of a British Commando'. The caption sneered that the much-vaunted 'Commando heroes' were frail after all and many more would be killed, as this 'ridiculously uniformed' Scotchman (sic) had been.

After the raid was over and the last of the British soldiers and seamen had been captured, the Germans were still suffering from shock at the British men's ability to penetrate their defences. However, to their great delight *Campbeltown* had not blown up. Admiral Kellermann, commandant of the St Nazaire base, went aboard with several officers and engineers to study the wreck. Having made this inspection, Kellermann departed but several of his officers remained, some of them with women friends. Close behind the cordon of troops ringing the ship many other Germans approached to gape at the enemy destroyer. Just before midday on 28 March the time pencils reached their detonators. The resulting explosion from the 4 tons of explosives, made even more devastating by the concrete packed around it, was cataclysmic. It killed 60 German officers and 320 soldiers. The caisson gate crumpled at the point of ramming, its ends protruded from their hinges and the inrush of sea forced it inside where it smashed against the west wall. *Campbeltown* had done its job.

Many British soldiers and seamen were fished out of the estuary by the Germans while others were washed ashore, astonishingly alive after many hours in the cold sea. Some, including Beattie – one of only two *Campbeltown* officers to survive – were allowed to rub down but were provided with nothing more than a blanket for the hard days and long journey before them.

The raid had reached its climax when *Campbeltown* blew up but there was yet an after-shock for the Germans. At four o'clock on the next afternoon, 29 March, torpedoes fired by the MTB at the Old Entrance exploded. The mystified and startled Germans were angry and immediately imposed a curfew with summary punishment for those who breached it. They feared that some raiders might still be alive and continuing their work. An hour later the second torpedo exploded. A group of French workmen, fearful of German suspicions and reprisals, dropped their tools and ran in panic. German troops fired at them and within minutes confusion and alarm seized the whole of St Nazaire. In the coming darkness the usually disciplined Germans attacked one another and their casualties mounted. Few if any of the raiders knew about these events until after the war but the French Resistance reported them to London.

Five Commandos, with daring, skill and some luck evaded capture and after many dangers reached Britain, to be taken back into the Army and to serve again during the invasion of Normandy more than two years later.

Not all German behaviour was atrocious in the aftermath of the raid and that of one man was outstanding in its humanity. This was Kapitänleutnant Paul, skipper of the destroyer *Jaguar*. A week after the raid Paul called on Colonel Newman in the prison camp at Rennes. In a formal but friendly way he said that he wished to bring to the colonel's notice the gallant conduct of a sergeant on Motor Launch 306. 'You may wish to recommend this man for a high award,' Paul said. He was referring to Sergeant Durrant, who was awarded the Victoria Cross. British officers had also witnessed Durrant's courage but perhaps Paul's report carried some weight with the VC assessors.

Frank Durrant's background and citation are worth noting. Born at Green Street Green, Farnborough, Hampshire, he had joined the Royal Engineers in February 1937 and in March 1940 took part in the expedition to Norway. Posted to an Independent Commando Company later that year he found it was service he liked. He had always been considered a 'born soldier'. His VC citation reads:

For great gallantry, skill and devotion to duty when in charge of a Lewis gun in HMS Motor Launch 306 in the St Nazaire raid.

Motor Launch 306 came under heavy fire while proceeding up the River Loire to the port. Sergeant Durrant, in his position abaft the bridge, where he had no cover or protection, engaged enemy gun positions and searchlights on shore. During this engagement he was severely wounded in the arm but refused to leave his gun.

The Motor Launch subsequently went down river and was attacked by a German destroyer [Paul's *Jaguar*] at 50–60 yards range and often closer. In this action Sergeant Durrant continued to fire at the destroyer's bridge with the greatest coolness and with complete disregard of the enemy's fire. The Motor Launch was illuminated by the enemy searchlight and Sergeant Durrant drew on himself the individual attention of enemy guns and was again wounded in many places. Despite these further wounds he stayed in his exposed position, still firing his gun, although after a time only able to support himself by holding onto the gun mounting.

After a running fight, the Commander of the German destroyer called on the Motor Launch to surrender. Sergeant Durrant's answer was another burst of fire at the destroyer's bridge, although now very weak he went on firing, using drums of ammunition as fast as they could be replaced. A renewed attack by the destroyer eventually silenced the fire of the Motor Launch but Sergeant Durrant refused to give up until the destroyer came alongside, grappled the Motor Launch and took prisoner those who remained alive. This very gallant Non-Commissioned Officer later died of the many wounds received in action. Sergeant Durrant's gallant fight was commended by the German officers boarding the Motor Launch.

The citation did not mention that Sergeant Durrant kept on fighting while wounded in both arms, both legs, stomach, chest and head. No wonder that the enemy officers were impressed. They did their best for him on the destroyer and sent him to hospital but with so many serious wounds there was no possibility of his surviving. He is buried in the Commonwealth War Graves Commission cemetery at Escoublac-la-Baule, near St Nazaire. The award of his VC was not gazetted until 19 June 1945 after the officers who were taken prisoner returned home to speak of his heroism and refusal to surrender and to confirm Kapitänleutnant Paul's report. Durrant's gallantry was similar to that of Able-Seaman Bill Savage, who also died on the River Loire during the raid and who also was awarded the VC. Three other VCs were awarded. They went to Lieutenant-Commander Beattie, Lieutenant-Colonel Newman and Commander R.E.D. Ryder. A few months after the raid the commandant of Beattie's Offlag ordered a ceremonial parade at which he read out the citation of the recently promulgated VC awarded to him 'as a tribute to the crew of the *Campbeltown*'.

For the sake of the historical record it should be said that Troop Sergeant-Major G.E. Haines of No. 2 Commando should have been awarded the VC. He was given the lesser Distinguished Conduct Medal but this really was not enough. Haines's sustained courage and example was witnessed by many of his comrades, including Lieutenant-Colonel Newman, who said of the sergeant-major, 'I cannot pick out all the chaps who made the breakout and the crossing [of the dock bridge] possible but outstanding among them was Troop Sergeant-Major Haines, who was superb. He alone knocked out several pockets of resistance with Tommy-gun and Bren-gun fire. He always seemed to have a fresh weapon in his hands.' Many of Haines's actions were on his own initiative and he was so prominent that German snipers and machine-gunners tried to pick him off. I know of several VC exploits which, though outstanding, were no more gallant than those of Sergeant-Major Haines.

A large number of gallantry awards were made to men who took part in Operation Chariot, more than for any other single operation of the war:

Victoria Cross	5
Distinguished Service Order	17
Military Cross	11
Conspicuous Gallantry Medal	4
Distinguished Conduct Medal	5
Distinguished Service Medal	24
Military Medal	15
Mentioned in Despatches	52

But these impressive decorations were not enough. Every officer and man who took part in the raid deserved at least the MiD – the Mention in Despatches. Some MiDs did go to men who were killed. At that time and for long after, the MiD was the only recognition that could be conferred on a dead serviceman, other than the VC itself, and it is probable that many of the officers and men 'mentioned' would have been awarded an actual decoration, perhaps even the VC, had they survived the action. Sergeant Durrant and A.B. Savage were killed in the performance of the exploit which resulted in the VC but their deeds were well attested at the time and VCs were predicted. We shall never know how many Army men would have been decorated with the DSO, MC, DCM or MM or how many Navy men might have been given the CGM, DSO, DSC or DSM.

I have mentioned nothing about the RAF in connection with the St Nazaire raid. This is because it did virtually nothing. The RAF was said, glibly, to have 'pressed the attack with determination' but only four aircraft actually dropped any bombs and then only one bomb per run. In England on the day after the raid, a squadron leader, feeling shamed and bitter, said, 'If only you had told us what it was all in aid of, we would have come down to nought feet and given them everything we had.' He meant both that the RAF would have supported the raiders at the risk of pilots' lives and that they would have hit the Germans hard. Winston Churchill said later, 'The only person you can blame for lack of air support is me.' As so often, he had insisted on being involved in planning that should have been left to the professionals.

One substantial, if abstract, benefit of the raid for the British was the reciprocal anger it provoked between the German Army and Navy chiefs. Both came under Hitler's lashing tongue because he insisted on finding somebody to blame for the British success. The Army's leaders put up the better defence, through Field Marshal von Rundstedt, Commander-in-Chief of all German forces in Europe, and General Jodl, Director of Operations. In consequence, Hitler apportioned the greater blame to the Navy. In this, at least, Hitler was right.

Admiral of the Fleet and Director of Combined Operations, Lord Louis Mountbatten, said of the St Nazaire raid:

Of all the operations with which I was concerned, the successful raid on the battleship dock at St Nazaire is the one I am most proud to have been associated with. It was one of those actions which can only be attempted precisely because it must appear to the enemy to be absolutely impossible to undertake. For this reason alone it required courage and determination of a quite unusual order to carry out. What is more, the complicated operation had been planned in greatest detail; and once the forces were committed each individual small ship and each little group of soldiers had to fight it out, guided by orders received at home, but entirely in the light of their own initiative. I know of no other case in military or naval annals of such effective damage being inflicted so swiftly with such economy of force. For in less than half an hour from the moment the *Campbeltown* rammed, all the Commandos' chief objectives were successfully achieved. This brilliant attack was carried out by night under a vicious enemy fire, by a handful of men, who achieved with certainty and precision what the heaviest bombing raid or naval bombardment might well have failed to do.

He might have mentioned that the St Nazaire dock was out of action for the rest of the war. He should have referred to the sacrifice. One hundred and fifty-nine officers and men were killed. There might have been 'economy of effort', as Mountbatten claimed; there was certainly no economy of sacrifice in this raid, which is considered by many to be 'the greatest raid of all'.

13

THE BRUNEVAL CAPER

OPERATION BITING, 27/8 FEBRUARY 1941

Early in the winter of 1940/1 the British High Command, and especially the RAF, was worried by the increased efficiency of German anti-aircraft gunners and of the night-fighter pilots. They seemed to be ready and waiting for the British bombers as they crossed the French coast and losses in men and aircraft were mounting.

The British were aware of the German early warning system known as Freya and the scientists had learned how to beat it. They did this by what was then sophisticated jamming and they could even mislead Freya. So how had the enemy improved their defence system? Various intelligence reports suggested a new form of radar.

This was the Würzburg. Made by the Telefunken engineers, Würzburg was not only accurate, it could 'read' the altitude of the British intruders, thus allowing the flak gunners time to set their sights to a nicety. Also, the night-fighters did not need to scour the skies for the RAF planes; the Würzburg operators instructed them to climb to, say, 10,000 feet and there they would certainly find their targets. Nobody on the British side knew exactly what the Würzburg looked like but reports from the French Resistance suggested that it incorporated a 'great dish' as an aerial. On 3 December 1940 a crisis meeting to discuss Würzburg was held at the Intelligence Wing of the Photographic Reconnaissance Unit (PRU) at Danesfield, Oxfordshire. The principal item on the agenda was a report about a supposed Würzburg which had been sighted at Cap d'Antifer, between Le Havre and Dieppe. Squadron Leader Anthony Hill, one of the PRU's best pilots, flew over Cap d'Antifer the next morning and took photographs which showed that radar equipment with a 20-foot diameter bowl or dish had been installed there. However, photographs could not possibly reveal how Würzburg worked; to do that the scientists needed to examine its components.

Lord Louis Mountbatten, who had succeeded Sir Roger Keyes as Chief of Combined Operations, authorised a raid by paratroops who would steal the Würzburg parts needed and return to Britain by sea. In 1941 this was a revolutionary idea. Only one minor raid had been made by British paratroops, in Italy, and all had been captured.

A company of the Parachute Regiment, a troop from No. 12 Commando and a detachment of Royal Engineers would combine to undertake the Bruneval mission, once French Resistance had reported on German defences at Bruneval and the terrain to be negotiated. While the Resistance was so engaged, command of the operation was given to Major John Frost, but the key technical figure was Flight-Sergeant Charles Cox, a leading radar expert. Only he would know what components of Würzburg to bring home. Another important member of the team was 'Private Newman', the *nom de guerre* of an anti-Nazi German who was to be Frost's interpreter.

Dr R.V. Jones, the Air Ministry's brilliant scientist, together with another radar specialist, briefed Frost, his officers and Sergeant Cox in the likely intricacies of Würzburg. Meanwhile, PRU developed their photographs of Bruneval and the Würzburg installation into a detailed scale model, which was studied by all officers and NCOs involved in the operation.

Frost's plan for the raid was straightforward. Group 1 of his paras would seize and secure the line of withdrawal. Group 2's job was to block any movement of German troops from their billets at Le Presbytere farm and elsewhere towards the Gothic villa, code-named 'Lone House', where the Würzburg was situated. Group 3, under Lieutenant Peter Young (later to become one of the greatest ever Commando raiders) would protect Würzburg while Cox did his work. From the same site, Major Frost planned to direct operations. Wing Commander Charles Pickard (later the hero of the RAF's raid on Amiens prison in 1944) led the converted Whitley bombers which dropped the paras.

The raid took the Germans by surprise but they quickly responded. One difficulty for Frost was that Lieutenant Charteris and his three assault sections had been dropped well away from the target area. Frost was without his full fighting strength. Sergeant Cox, aided by Lieutenant R. Vernon, was soon at work on the Würzburg. After the meticulous planning and briefing it came as a shock to the pair to find that they had no screwdriver long enough to remove the vital pulse unit. With time running short they broke it off with a crowbar.

Frost urged more haste for German resistance was stiffer by the second. A company of infantry who had just returned to their billets in the hamlet of La Poterie was heading at the double towards the Bruneval cliff top. To Frost's relief, Lieutenant Charteris and his men now arrived, after a run of 3 miles, and they joined the operation to safeguard the stolen equipment and protect the withdrawal route.

Soon after 1 a.m. Cox and Vernon loaded the Würzburg parts onto a wheeled trolley specially built for the snatch and hurried down a cliff path to the evacuation beach. After anxious delays caused by faulty signalling equipment, the Navy received Frost's call to come in and make the pick-up. Eight infantry landing craft approached right to the water's edge. The raiders scrambled aboard and the LCIs moved off, to be taken in tow by gunboats. The protection provided across the English Channel was impressive – air cover by Spitfires and escort by two British destroyers and some ships of the Free French Navy.

Within a few hours of its arrival in Britain, the Würzburg was being studied by Dr Jones and his scientific team. A direct result was the invention of 'window' – strips of metal foil which RAF bombers dropped to confuse the enemy radar operators.

Operation Biting had been an outstanding raid. Major Frost lost two men killed, two wounded and six taken prisoner. The Germans lost five killed, two wounded and two taken prisoner. Adolf Hitler was furious and caustically criticised his intelligence chiefs and those generals in command on the French coast. It seemed to him impossible for the British to have pulled off such a coup without betrayal by his own people. There had been no betrayal but the work of the French Resistance in carrying out their detailed ground reconnaissance had been vital for the success of the Bruneval caper.

The Germans had formed a unit, the Brandenburg Division, for special operations and it was remarkably successful but it never managed to seize a British radar installation in reprisal, as Hitler demanded.

The Bruneval Caper demonstrated the need for reliable information from people on the ground in the region where a raid was to take place. Ideally, British agents – who did not necessarily have to be British subjects – should be inserted but this was not always possible and sometimes it could not be done in time.

During the preparations for Operation Biting the British planners had the assistance of a very special French Resistance man, Gilbert Renault, known at the time only as 'Remy'. Remy's *reseau* or resistance cell was the Confrérie Notre Dame, known simply as CND in British records. CND was always busy and Remy had a reputation for accurate and reliable information. Among his projects at the time was a constant watch on the German battleships *Scharnhorst* and *Gneisenau*. They had spent almost a year in the port of Brest and Remy suspected that they were getting ready to make a dash for the better safety of German ports.

On 24 January 1942 an urgent signal reached him from London. His British colleagues wanted information on five matters:

- How many machine-guns defending the road at Theuville between Cap d'Antifer and Jouin?
- What other defences are in that region?
- How many enemy are in position, what units are they from and what reputation do they have?
- Precisely where are the German troops quartered or in huts?
- Provide sketches showing positions of barbed wire.

This was a comprehensive assignment and the experienced Remy realised that something important was being planned. Remy's own significance lay largely in his ability to find the best men or women for an intelligence mission. He chose 'Pol' and 'Charlemagne', who went to work at once. Having booked themselves into a hotel at Le Havre, posing as agricultural engineers, they then drove to Bruneval and spoke with the friendly and talkative proprietor of the Beauminet, an hotel-bistro which turned out to be used frequently by the German troops. Was it safe for them to move about, Pol and Charlemagne asked. The proprietor asked the Germans who were actually quartered in his hotel. They obligingly warned the two anxious Frenchmen that they should not go close to certain places because machine-guns were stationed there. Nevertheless, equipped with a local map showing the various farms which the 'agricultural engineers' said they proposed to visit in the course of their work, Pol and Charlemagne occasionally went right up to a barbed-wire trestle-barrier across a road. They asked the young German sentry if they could take the path down to the sea. Better not, he said, the beach was mined. However, the two agents concluded that the mines had not yet been laid.

They set eyes on the senior sergeant who commanded the Bruneval garrison and judged him to be keen and competent. Apparently he was absent at times – visiting friends in nearby units. Pol and Charlemagne had little trouble finding out which units, because many soldiers had signed their names and regiments in the hotel visitors' book. The Frenchmen did not steal the book for that could have been suspicious but they copied from it while pretending to read agricultural equipment brochures. In one way or another the agents found out al that London needed to know – and much more. Remy saw to its despatch.

Major Frost and his raiders would never have known where all this helpful information came from. But not all raiders, wherever they operated, were assisted by the quality of the intelligence provided by Remy's CND.

14

'PINPRICK RAIDS' ON THE FRENCH COAST, 1942

The Würzburg at Bruneval was just one of many targets along the French coast, which was the natural arena for the early raids, whether carried out by Commandos or other raiders.

For some rugged individualists even a Commando company or troop was too large. Needing more independence, these men formed their own small outfits. Perhaps the smallest group was that known as Small-Scale Raiding Force (SSRF), which was formed by Major Gus March-Phillips DSO, MBE, Captain J.G. Appleyard MC and Lieutenant Graham Hayes MC. It is unclear how they managed to obtain official approval for the establishment of their tiny unit, which was also known as No. 62 Commando, but it must have had a lot to do with March-Phillips's driving personality and his political contacts.

Whatever the circumstances of its foundation, SSRF's mentor was a Commando himself, Brigadier Robert Laycock, who was then commanding a Special Service (Commando) Brigade. Earlier, Laycock had raised No. 8 Commando and in a very real sense he was one of the founding fathers of the Commando organisation. Any officers with a scheme to 'get at Jerry' could be sure of Laycock's support.

To call SSRF a 'force' was an impudence but the word had an agreeable ring in 1942. The three venturers set up business in Anderson House, an Elizabethan manor near Poole Harbour, Dorset. Its owner, John Stevenson, stayed on at the house as host. There was no approval for this irregularity but it was in keeping with the spirit of the SSRF.

These spirited and adventurous men mounted their first expedition on 14/15 August 1942 when they set out to destroy a German anti-aircraft gun at Cap Barfleur. They reached the enemy coast in an MTB, launched their Goatley collapsible canoe and paddled in to the shore. The MTB commander had landed them at the wrong place so they did not even locate their target gun. Not wishing to return to base entirely unsuccessful, SSRF killed three German soldiers – killing enemy troops was always considered something for satisfaction and praise and for congratulations as well, even after it became commonplace. A 'day trip' to France with the object of committing legalised murder might seem strange to modern readers but it was considered absolutely normal in 1942.

The operation on the night of 2/3 September was more successful. Appleyard led a raid on the Casquets lighthouse in the sea off Cherbourg, which the Germans had been using as a naval signalling station since 1940. The raiders' task was difficult because of the violent tidal races around the rocks which the lighthouse guarded. Appleyard was navigator for the small LCI and it was his job to jump for the rocks, taking with him a light line. He held the LCI up to the rock while Hayes, in the stern, kept the boat off the rocks with a line and anchor which he had dropped as the LCI approached. With the entire party of twelve Commandos ashore, Appleyard left the LCI in Hayes's charge, with a man to manage the bow. He led the others through barbed wire and crossed the courtyard without challenge. Signalling the men to go about their pre-arranged tasks, Appleyard and Sergeant-Major Tom Winter ran up the spiral staircase to the light room, which was empty. The garrison was in another room and they were stupefied when the Commandos burst in. They had small arms and grenades at hand and an anti-aircraft pom-pom gun was set for use but not a shot was fired.

The Commandos dumped the weapons and ammunition into the sea from the lighthouse, smashed the searchlight mechanism and seized some code books, which were to prove invaluable. They captured seven Germans, three of whom were in bed. Another two were turning in and two others were going about their routine tasks. Shepherding these prisoners before them, the Commandos returned to the LCI. Appleyard, who broke a bone in his leg while re-embarking, was the only casualty.

When they returned to Portsmouth they were told by a communications watch station that the German HQ in Cherbourg had been frantically calling up the Casquets lighthouse by radio. The German crew of the lighthouse were probably better off as British POWs than they would have been if Appleyard had left them behind to explain what had happened. Hitler could have vented his full fury against them and had them shot.

Three nights later March-Phillips led a raid on St Honorine in which Appleyard and Hayes also had roles to play. Because of his damaged leg Appleyard was navigator of the MTB while March-

Phillips, Hayes, Sergeant-Major Winter and eight others formed the raiding party. March-Phillips found that the objective, an enemy ammunition depot, was all too heavily guarded and he withdrew his patrol, planning to make a return visit with a larger force.

About 200 yards from the beach the raiders heard a German patrol, which they ambushed, killing seven men. March-Phillips was searching the bodies for maps and any other valuable intelligence material when a larger party was heard approaching. March-Phillips doubled his men back to the assault boat, which they jumped into and paddled furiously. They were too late. A flare went up and the Germans opened up with a machine-gun. They killed March-Phillips and three men and sank the boat. Hayes had not been hit and as he swam away he shouted to Appleyard that it was hopeless for him to remain there with the MTB – he should save his ship. But Appleyard's first priority was to search for survivors and before he could pull away the MTB was hit and lost an engine.

Meanwhile Hayes had swum some way along the coast and when he came ashore he was able to evade the Germans. He travelled to Spain on the Resistance routes but here the pro-Axis Spanish handed him over to the Germans. They shot him in Fresnes Prison, France, on 13 July 1943. He was another victim of Hitler's infamous 'Commando Order', that any Commando taken prisoner was to be killed. A French guide who had been with the SSRF patrol also reached Spain, where the police tortured him. He was virtually disabled by this cruel treatment but he escaped and returned to England. He volunteered to go back to France as an agent and was very successful.

Dramatic adventures befell Sergeant-Major Tom Winter. Having swum unseen close to the MTB he was profoundly disappointed when Appleyard had to retire. When he struggled back to a beach a German attempted to shoot him but missed. Captured, Winter was taken to a prison camp in Poland. This resourceful man found that he could get out of the camp at night and back in before dawn. Making contact with the Polish underground, he gave them lessons each night on the use of explosives. The authorities suspected him after a time and he was sentenced to 10 years' solitary confinement. He was liberated by the Russian advance and then marched westwards to rejoin the British Army.

With his friends dead or believed to be dead, Geoffrey Appleyard had more reason than ever to kill and discomfort the Germans and on the night of 3/4 October 1943 he was raiding again. He led four officers and five men to the island of Sark on an intelligence-gathering patrol. They captured five Germans who were in bed at the Dixcart Hotel and unwittingly began a series of events which were to have dire repercussions. One of the Germans was an amazingly strong man who could not be controlled. Even at the point of a revolver he was difficult to handle so his hands were tied. Perhaps Appleyard did not realise that this was against international law regarding prisoners. However, when the Germans saw newspaper photos of their man so constrained they understood its implications at once. Commando prisoners earlier taken at St Nazaire and Dieppe were handcuffed as a reprisal.

These small raids and others by SSRF men were mere pinpricks but they achieved the aims of convincing the French that the British were still militarily active, of unsettling the Germans and of gaining experience for future operations.

Operation Musketoon was a more ambitious project and although it had a tragic end it achieved more for the British war effort than the 'pinprick raids' of March-Phillips, Appleyard and Hayes. On 2 November 1942 Captains Gordon Black and Joe Houghton of No. 2 Commando led a party of Commandos and Norwegian Resistance men in an attack on the hydro-electric power station at Glamfiord, Norway. This plant provided all the current for the main aluminium factory in Norway, a metal vitally important for Germany in the building of aeroplanes and vehicles.

The Commandos landed from a Free French submarine and approached their objective in the dark across a dangerous glacier of black ice. Just before midnight, still unseen, Black and Houghton and their men broke into the plant. In the process one German sentry was killed. The Norwegian workers were locked up so that they would be unable to raise the alarm for several hours. The Commandos then went about their business of wrecking the machinery with explosives and sledgehammers. They even destroyed some of the pipelines. A German patrol quite by chance encountered the Commando party and in the wintry night a fire fight ensued. Houghton and Black were wounded and captured. As a consequence of Hitler's order the previous month that captured British Commandos should be butchered the officers and their men were taken to Germany and shot.

The death of March-Phillips and Graham Hayes was not the end of SSRF. As No. 62 Commando it grew in size and in September 1943 Bill Stirling, David Stirling's brother, was appointed to command it. Geoffrey Appleyard was a member of No. 62 Commando at this time and one of its most experienced officers; he was killed in North Africa in 1943 while operating with an SAS unit.

The loss of Gus March-Phillips in the St Honorine raid was more than a personal tragedy. He had achieved much even before he co-founded SSRF and he would have gone on to raiding greatness. Like the others, he had known the risks but was prepared to face them for the sake of defeating the Nazis.

About 18 months earlier March-Phillips had led a raiding party to Nigeria where, at Lagos, they sank two German merchantmen and stole a third. On the return journey March-Phillips noticed that one of his men was a mere lance-corporal while all the others were officers. The odd man out was Anders Lassen, a Dane, who would go on to fulfil March-Phillips's prediction that he would become one of the war's best raiders (see Chapter Seven). Calling Lassen to him during the voyage back to Britain, March-Phillips said, 'Andy, I think we had better have you commissioned'. This was how Lassen got his first pip, as a second-lieutenant. At the end of the war he was Major Lassen with the Victoria Cross and three Military Crosses to his name. His was a case of death and glory for he was killed during the exploit that resulted in the VC award. Gus March-Phillips would have been proud of his man's career.

15
PACIFIC RAIDERS

CARLSON AT MAKIN ATOLL, AUGUST 1942

After making their vigorous and irresistible sweep across great swathes of the Pacific Ocean early in 1942, the Japanese garrisoned many coral atolls as part of a strategy to prevent the Allies from making a grand counter-offensive. One of them was an apparently insignificant scrap of coral, mangrove swamp, sand and palm trees around a lagoon called Makin.

Commanding the garrison was a formidable sergeant-major named Kanemitsu, with about 250 men under his charge. The very fact that a sergeant-major rather than an officer was responsible for Makin was proof enough that the High Command in Tokyo did not consider the place important.

However, Makin soon assumed a significance to the Americans following the landing of 20,000 Marines, under General Alexander Vandegrift, on the major position of Guadalcanal in the Solomon Islands on 7 August. The Japanese fiercely contested possession of Guadalcanal and it was imperative to prevent them from reinforcing their troops there. One way for the Americans to do this was to attack other islands and draw reinforcements to these places so that they would not be sent to Guadalcanal. Makin was not in itself important in US strategy but it would be sufficient if the Japanese could be deceived into thinking that it was and tie up their resources on the island. Thus, Lieutenant-Colonel Evans Carlson was ordered to capture the place with his 2nd Raider Battalion. It was emphasised to Carlson that the operation was indeed to be a raid and he was instructed to complete it in a day. A brash extrovert, Carlson was just the man for the job.

Carlson believed in total personal command and control and it was his habit to carry around a portable public address system. This was generally on the back of his signaller while Carlson carried the loud-hailer. His voice boomed over the training grounds and his area of competence in battle. His war cry and that of his battalion was 'Gung ho!' Carlson's force consisted of 13 officers and 208 other ranks and it departed from Pearl Harbor, Hawaii, on 8 August in the submarines *Argonaut* and *Nautilus*. The trip to Makin was peaceful and the subs travelled mostly on the surface. Inspecting the atoll through a periscope, neither Carlson nor the naval officers saw any enemy movements and all stayed quiet as the score of rubber boats, powered by outboard engines, were offloaded from the subs.

Then came Carlson's first problem: the breaking surf with one giant roller rapidly succeeding another was unbelievably noisy. Even with his loud-hailer the colonel could not make himself heard. He was command-impotent. The sea was rough, currents were running crazily and the two waves of invaders, heading as planned for different beaches, could not keep to their original landing intentions. Only fifteen boats reached the shore. Still, nature's noise concealed the raiders' approach and they might well have overwhelmed the defenders by surprise had not one Marine accidentally discharged his rifle. Not for the first time – or the last – inadequate firearms training destroyed the element of surprise. Sergeant-Major Kanemitsu heard the shot and quickly prepared his garrison to repel the American assault.

A capable leader, Kanemitsu already had machine-gun posts strategically positioned and his snipers at once ran to their allotted palm trees, climbed them with an agility unmatched by any Western soldiers, and hid among the fronds. Kanemitsu sent off a radio message to his superiors on Jaluit Island and as a result two ships in the area were diverted to Makin. Kanemitsu did not know about the US submarines waiting below the surf until they opened fire with their 6-inch guns and sank both ships. Not one of the seventy or so soldiers aboard them survived. Most were picked off as they swam towards the shore.

But this did not affect Kanemitsu's defence and within minutes of the American advance starting, it lost momentum under Japanese fire. At this critical point Sergeant C. Thomason took charge and in several individual exploits he demonstrated to his marines that the Japanese, far from being super-soldiers, were only human and could be beaten. At one point, in full view of his men he boldly walked up to a house in which an enemy sniper was concealed, forced open the door and shot the man before he could resist. Thomason also led a counter-attack against positions at the lagoon and broke the Japanese marines' will to resist. Thomason lost his life in yet another fine act of leadership that day, being shot as he encouraged his men forward. He was awarded the Congressional Medal of Honor, whose citation lauded his bravery and splendid example.

The raiders' Company A occupied Government House and other installations but could not link with Company B. They were still separated at noon when Kanemitsu's emergency call was further answered – this time by twelve aircraft, including six light bombers. They bombed the raiders for about an hour, partly to safeguard the landing on the lagoon of two large troop-carrying flying boats of the Japanese naval air corps. One was set on fire within minutes of its landing; the other was shot to pieces when the alarmed pilot attempted to take off before discharging his soldiers. Most of these reinforcements survived the destruction of their flying boats but all were shot dead as they tried to reach shore. At no time did any Japanese attempt to surrender – and they were not invited to do so: raiders did not take prisoners unless they were needed for intelligence purposes.

Snipers in their palm trees were now the greatest threat to the Americans, especially those of Company A. It was difficult for the raiders to pinpoint a sniper's position and casualties were mounting. At this point Carlson showed that for all his flamboyance he was as good an opportunist raider leader as any of the war. He spotted more enemy aircraft approaching the Company A position and saw the chance to exploit them. He ordered a junior officer to protect the flank of the American position with his platoon, the others he commanded – this time through his loud-hailer – to withdraw rapidly and at once to a fresh position more than 200 yards away. His men may have thought him crazy to do this in the teeth of an air raid but Carlson was taking a calculated risk. As the Americans ran away, the snipers climbed down from their trees to give chase – just as their own planes screamed into the attack with machine-guns blazing and aerial bombs exploding among the Japanese. That took care of the snipers just as Carlson had anticipated. No American was wounded by the planes.

Kanemitsu and his men were doughty fighters and they made three attacks against the entrenched Americans during the day. All were beaten back but Carlson knew that during the night the garrison would be reinforced. He made an orderly withdrawal to the beach ready to be taken off but again the heavy surf was his enemy. Boats capsized and wounded men in them drowned. Only a few raiders mostly strong swimmers, reached the waiting subs.

About midnight Carlson calmly told his men that any who felt like it could try to reach the submarines, though the risk was great. He himself proposed to lead the others to another part of Makin where he now knew there was a village with outrigger canoes. Using these craft he hoped to reach *Nautilus* or *Argonaut*. It was the fate of the wounded that concerned Carlson throughout the action for he knew that the Japanese would butcher them.

During the night about seventy men in rubber boats and on makeshift rafts reached the subs. Everybody knew that Carlson would stay as long as he had wounded to care for and possibly evacuate. The *Nautilus'* commander sent off towards the shore one of the few rubber boats with its engine still working. From this boat a man swam to the beach with a line which could be used to haul the damaged boats through the surf. But now it was broad daylight and the subs and the raiders were vulnerable to renewed air attack. A plane shot up the boat carrying the line and all the men in it perished. The naval commander signalled that the subs would return that night and then they crash-dived.

That day Carlson led a party to find out what the Japanese were doing and, perhaps even more important, to locate food and drinkable water. The foraging was successful but there was no sign of the Japanese until one of Carlson's officers rowed out to inspect an anchored sloop. As the officer and his men were making fast to the sloop a Japanese fired a pistol at them through a porthole. The raider lieutenant responded by dropping a grenade through the porthole. This was a bad mistake because, while the explosion killed the Japanese it also sank the boat. Carlson could be forgiven his anger. He could have used the sloop to ferry his men to safety.

During the long hours as they waited for night to fall and the submarines to return, Carlson had time to think of his twenty-one dead. He paid the Makinese islanders to give them a proper Christian burial – and this was done. Then he had his remaining boats and an outrigger canoe lashed together. With only one boat's engine working they set off to join the submarines and after an exhausting and anxious 3 hours they did so. Carlson had done well and his small operation had indeed drawn some pressure off Guadalcanal.

He might have been satisfied, but that was before he found that nine of his men were unaccounted for. He was criticised for this but it was not really his fault. The men had set off in a rubber boat but the pounding waves tore the paddles from their grasp and they drifted helplessly. The Japanese returned to Makin in force the day after the raiders left, captured the nine Marines and took them to one of their bases. After interrogation they were beheaded on the orders of Vice-Admiral Koso Abe. The execution was witnessed by islanders who four years later gave evidence before a war crimes court. Abe was executed after the trial and the officer who carried out Abe's order was gaoled for 10 years.

As an individual commander Carlson showed yet again that raiders are best led from the front. He claimed a great victory and because his dramatic enthusiasm was infectious, the press lauded his operation. The raiders, according to one newspaper report, were 'experts in death, demolition and destruction'. Another paper published the raiders' battle song, which was sung to the tune of '*Ivan Skavinsky Skavar*', though nobody sang it on Makin. Carlson was a dashing leader, reporters wrote, and wherever he went, shouting 'Gung ho!' his men would follow. His chiefs did not quibble about these press assessments of Carlson and his battle, even though there were disturbing rumours that he had proposed surrender. Carlson's glory reflected on the US Marine.

Much later the opinions of the official historian were more realistic. He stated that the significance of Carlson's raid was negligible. This was true enough but the historian, a Navy man himself, obeyed the close-ranks policy so common in the US services and did not even mention the nine Marines who had been left behind or Carlson's surrender attempt. There was a practical lesson in the Makin raid – paddling rubber boats in an island surf was suicidal – but surely this should have been obvious before the operation even began.

The Makin raid was a mere footnote to the titanic struggle that engulfed the western and southern Pacific but it proved that the Americans could mount and sustain a raid. Whether in this case it was worth the thirty Americans dead and fifty wounded is open to debate.

16

COCKLESHELL HEROES

ROYAL MARINE BOOM PATROL DETACHMENT (RMBPD)

If ever an outfit of raiders suffered from being given an apparently meaningless title it was the RMBPD. A boom was the protective steel underwater 'fence', suspended from buoys, which was placed across the mouths of rivers and estuaries to keep out and hopefully trap enemy underwater or surface raiders. Britain had booms right around its shores. A cover name was needed by the Royal Marines who formed the detachment and as they trained near the Solent boom in Hampshire it was suggested that 'boom' be incorporated into their unit designation. The cover story, for the curious, was that these Marines were guarding the Solent boom.

The RMBPD's duties were nothing like as defensive as their label might have suggested. These were aggressive warriors and their job, using small craft of various kinds, was to penetrate through German defences and attack major targets. The men who carried out the most famous of all RMBPD missions became known as 'the Cockleshell Heroes', after the Folboat Cockle Mk II, in which they did their raiding.

The two officers most responsible for RMBPD successes were Captain T.A. Hussey RN, who operated from the Combined Operations Development Centre (CODC), and Major H.G. 'Blondie' Hasler, who was the leader in the field. Hasler was a natural small-boats man. As a boy and youth he had much adventurous experience in canoes, sailing boats and yachts. During the campaign in Norway in 1940 he had commanded the Royal Marine Landing Craft Company (RMLCA). Years ahead of his RN and RM superiors in the quality of his imagination and ability to solve problems, in 1941 Hasler submitted to Combined Operations HQ a detailed plan for attacking enemy ships by the use of canoes and underwater swimmers.

It hardly needs to be said that his suggestions were put aside, by everybody, that is, except Captain Hussey. Every new idea in Britain was quashed until some catastrophe or crisis brought it back to the minds of the original doubters. In Hasler's case it was the Italian 'human torpedo' attack on British ships in Alexandria harbour in 1940. CODC now asked Hasler to give 'serious attention' to this new type of warfare. It was not new to Hasler, who had reasoned that he could not be the only man in the world with bright ideas. Intelligence had picked up rumours of 'developments' in the Italian Navy and Hasler was not surprised by their raid on British ships in harbour. After all, they had used explosive boats in the First World War. The British had one of them in a museum and Hasler had studied it. He did not copy it but it influenced his thinking.

With the ingenious marine designer Fred Goatley, Hasler produced a sturdy two-man canoe which was collapsible and could carry 150 lb of explosives and equipment. It had virtually no profile in the water and tests showed that if the crew wore a certain type of camouflage wet suit they were undetectable. Using double-bladed paddles men could move rapidly, but when these paddles, which had a high rise, might catch a reflection they could be converted in seconds to a single-blade, low-rise paddle.

Obviously, the canoeists needed a back-up, support and logistics team but having asked Hasler to give 'serious attention' to his project his superiors baulked at the idea of a new unit. It was Lord Mountbatten, as Chief of Combined Operations, who cut through the red tape. It was he too who wanted the title of the outfit to be changed from Royal Marines Harbour Patrol Detachment to Royal Marines Boom Patrol Detachment. The use of the word harbour would certainly excite enemy intelligence.

At Mountbatten's request, Hasler was appointed commander of the RMBPD with Captain J.D. Stewart as his deputy. As always with 'special and secret' units, volunteers were called for. There was no shortage of them even after the Marines were told of the basic requirements: they had to be keen on genuine active service, they must be strong men able to swim well and, oh yes, if they were concerned about their personal safety and anxious about a wife and children at home, they should not apply. The officers who did the recruiting, including Hasler, were also looking for men of great common sense. The one big advantage for the men was that, like the Commandos, the RMBPD Marines were allowed to find their own billets and they were paid Commando-rate subsistence.

Raising a force from scratch, Hasler was able to train it as he wished. With the help of Captain Stewart and his two hand-picked juniors, Lieutenants J.W. McKinnon and W. Pritchard-Gordon, he taught the Marines about navigation, explosives, underwater swimming and the qualities – and faults – of the various boats in which they would practise. Parachuting was part of the syllabus as was knife fighting and unarmed combat. Finally there came survival techniques.

Hasler said that loading men and their gear into canoes on the open decks of submarines took too long and was dangerous for two reasons. The first was that an enemy patrol boat or aircraft might spot the stationary sub engaged in the launching, the other was that in rough water a crew member might be washed overboard while getting into the canoe. Hasler insisted that the submarine's crew would launch the canoe with the two-man crew already aboard with all their stores and the cockpit covers fastened. There was some demurring about this because no such launching had ever taken place. Patiently, Hasler overcame the Navy's objections by designing a special sling, together with a collapsible derrick to put the canoes into the water.

The RMBPD's first operation, code-named Frankton, was immensely ambitious and daring. Twelve men in six Cockles would paddle up the great French River Gironde for 75 miles and mine enemy shipping, particularly those ships which were running the British blockade along the French coast – they were carrying war *matériel* to and from Japan. There was nothing new about the British desire for a raid on the Gironde ports, including Bordeaux, but air attacks and Commando raids and assault by RN warships were considered wasteful in life, not least of all French civilian lives.

Hasler had no doubt that he and his men could carry out their mission but getting back was a more serious matter. They would have to travel overland to Spain and hope to be airlifted home. The only other way of surviving was to be taken prisoner, but given Hitler's hatred for any type of Commando they could be executed on the spot.

The adventurers were Hasler, Lieutenant McKinnon, Sergeant Wallace, Corporals Laver and Sheard and Marines Sparks, Conway, Mills, Ellery, Ewart, Moffat and Fisher – two men to each Cockle.

Off the French coast they disembarked from their sub on 7 December 1942. Within a few hours several disasters occurred. The canoe carrying Ellery and Fisher was damaged when it fouled the torpedo hatch and could not be launched. The Cockle of Sergeant Wallace and Marine Ewart disappeared while the men were negotiating the violent and dangerous tide-race over shoals that had not been shown on the charts. Sheard and Moffat capsized in another tide-race. Hasler and Sparks rescued them and towed them to the bank. Here Hasler told Corporal Sheard that he and Moffat would have to manage as best they could. He himself and Sparks must continue with their mission. Sheard said, 'That's all right, sir,

I understand. Thanks for bringing us so far.' Like all compassionate raider leaders, Hasler was concerned for his men and upset by the prospect of their deaths. It is easy to imagine his feelings when parting from Moffat and Sheard; he knew that their chances of survival were minute.

Unknown to Hasler, Wallace and Ewart were soon captured but under rough interrogation they insisted that they were naval men who had been washed overboard from their ship. Their secret activities they kept to themselves. Hasler could only hope that McKinnon and Conway were safe and carrying out the predetermined plans.

Hasler, Sparks, Laver and Mills found a hiding place and there Hasler told them that while he and Sparks pressed on into Bordeaux, Laver and Mills would scout the eastern docks. If no better target presented itself these two would attack the two enemy ships at anchor which they had spotted while hiding.

Hasler and his mate then fixed three limpet mines onto a cargo ship, two on a small naval vessel and one on a tanker. Laver and Mills, forced to put their secondary plan into operation, attached five limpets to one ship and two to another. Quite by chance Laver and Mills encountered their commander and Sparks, and Hasler told them that the best chance for all of them was to split up; he knew that four obvious foreigners would soon attract strong suspicion. Laver and Mills were picked up, as were McKinnon and Conway. Sheard and Moffat were already in enemy hands. All six were shot. Hasler and Sparks, after incredible hardships and many misadventures, reached Britain in April 1943, five months after they had set off in their Cockle. In terms of destruction the raid had been successful. The first limpet exploded at 7 o'clock on 12 December and, at irregular intervals, they went on exploding. One cargo ship was sunk and four others and the naval ship were seriously damaged.

The Germans were dismayed at the apparent ease with which their defences had been pierced. Such was the degree of panic following Frankton and other coups that Admiral Donitz publicly warned his navy that an Allied operation was imminent. Adolf Hitler was, as always, furious. By this time one of the Cockles had been found and Hitler sarcastically demanded of his senior officers, 'How could this child's boat evade our supposedly impregnable defences?' The short answer to this question was that the 'children's boats' had been crewed by professional raiders – courageous and determined men. Hasler was decorated with the DSO, Sparks with the DSM. Many people believed that the awards should have been the VC, the decoration which was later awarded to naval Lieutenants Cameron and Place after their successful attack on the German battleship *Tirpitz* on 22 September 1943, and to Lieutenant Fraser and Leading Seaman Magennis for mining a Japanese ship on 31 July 1945.

In truth, it is difficult to see any qualitative difference between the Gironde exploit by RMBPD and the later missions. All were successful, all heroic. With his 'Cockleshell Heroes', Hasler – who later operated in the Far East – showed that no enemy target was invulnerable against attack by brave raiders.

17

DIEPPE, 18/19 AUGUST 1942

OPERATION JUBILEE

Dieppe was the biggest raid carried out by the British during the Second World War – though the Canadians provided the largest infantry contingent. I consider it a misbegotten operation, ill-conceived, poorly planned and lacking in a proper intelligence appreciation. A complete and tragic failure, it was relieved only by extraordinary heroism, determination and splendid examples of junior leadership.

In counter-balance, I must admit that the late Brigadier Peter Young, an old friend of mine and one of Britain's most brilliant fighting soldiers, told me that Dieppe was 'necessary'. He played a significant part in the raid – indeed one of the only two successful parts – and his opinion must be heeded. He said that because Dieppe showed the world that Britain still had aggressive intentions against the Germans this was justification enough for the raid. Further, he said, it had been necessary to study the conditions likely to prevail when the Second Front was launched.

'But Peter', I protested, 'was it worth the great cost in British and Canadian lives? And surely it had the unwanted effect of raising the morale of the German armed forces while damaging that of the British and Canadians. The world became aware of Britain's aggressive intentions, but even more publicity was given to the British failure.'

Peter accepted my points but in the end we could only agree to disagree; I include his comments in order to show that, as with many operations of war, there are opposing points of view.

Dieppe was chosen largely because it was within range of British fighter cover. Another reason given for the choice was that the target area was outside any stretch of French coastline where the great Allied landings were likely to be made one day. But it had a great defect from the British perspective: east and west of Dieppe port are high, steep chalk cliffs. They were a natural German defence and it must have been obvious to all commanders that Dieppe would be a tough nut to crack.

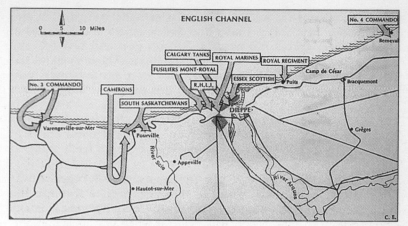

The British–Canadian assault plan for the raid on German positions at Dieppe.

Still, first-class troops had been chosen for the task. Dieppe itself was to be captured by six infantry battalions and an armoured regiment of the 2nd Canadian Infantry Division. At Berneval and Varengeville the Germans had installed coast defence batteries, so sited that they could take any ships approaching in cross-fire. These guns, the planners decided, had to be destroyed before the main attack on Dieppe itself went in. No. 3 Commando (Lieutenant-Colonel Durnford-Slater) would attack Berneval, while No 4 Commando (under Lieutenant-Colonel the Lord Lovat) dealt with Varengeville's guns.

Active planning began in April so it cannot be said that the operation was too hastily prepared. Indeed, it may have taken too long. In country similar to Dieppe but in the south of England – Lulworth Cove, Dorset – Lovat's Commandos eight times rehearsed their landing, in full gear and with exactly the same weight of equipment that they would carry during the assault. The briefings were impressively thorough.

The real operation began in August. At Varengeville, No. 4 Commando soldiers under Major Derek Mills-Roberts began well, reaching the top of the cliffs through a cleft that had been pinpointed by aerial photography. By the time they reached the top the six German heavy guns were already firing on the approaching landing convoy. The Commandos, still unseen, took up positions at the edge of a wood a mere 150 yards from the guns and with the crews in their rifle sights. Several Germans were hit in the Commandos' first volley. Under this fire, gunners could not serve their artillery pieces but soon a 20 mm gun and seven heavy machine-guns opened on the British positions. These weapons did not deter the Commandos. Mortars fired from behind cover were more dangerous, but two could play at this game and with a 2-inch mortar a Commando crew scored a hit on a stack of cordite near one of the enemy big guns. The fire from both sides was furious and Commando casualties mounted. A very gallant medical sergeant, Trevor Garthwaite, was mortally wounded while going to the assistance of an injured man.

At this point the flare signal for the general assault began. It was costly from the beginning as the raiders struggled through the heavy beach wire while under mortar attack. Commando F Troop, one of two with Lord Lovat, was held up when first its lieutenant-commander and then its sergeant were killed. Captain Pat Porteous, who was with Lovat's small HQ, ran across to F Troop, took command and prepared to lead them in a charge. A German attacked him and shot him through the wrist but Porteous killed his opponent with his other hand.

As B Troop attacked the Varengeville battery they heard a shot and saw a German emerge from a barn and crash his boots into the face of a wounded Commando. A soldier shot the enemy in the pit of the stomach, perhaps to give him a touch of the agony he had inflicted on the helpless Commando. 'We doubled across the yard to where the two wounded lay side by side,' said Corporal J.R. Gilchrist. 'For our comrade, morphine; for the beast, a bayonet thrust.'

Despite his wound Porteous led his men of F Troop to clear the enemy battery buildings and captured the gun sites. He was now wounded in the thigh but he was still the first man into the guns, leading his Commandos in a wild bayonet charge. He captured each gun pit in turn. Throughout he was closely supported by Troop Sergeant-Major Portman. They were seen to kill all but one of a German gun crew before they charged the next gun pit and seized it. A German grenade blew off the whole of Portman's right heel so he sat on the ground and carefully picked off Germans with his rifle.

With the Varengeville battery demolished and burnt, Lord Lovat withdrew No. 4 Commando in good order, though 2 officers and 10 men had been killed and 20 wounded. At least 150 Germans were killed. This part of the Dieppe attack, managed with peerless efficiency and Commando thoroughness, was a praiseworthy success. The assault on Berneval battery by No. 3 Commando, under Lieutenant-

Colonel Durnford-Slater, did not go as planned. Luck is always an element in a successful raid and Durnford-Slater's luck, which was generally good, was out on this occasion. He was to land on Yellow Beach with half the Commando while his second-in-command, Major Peter Young, would go ashore on Yellow Beach 2. This scheme was upset because Durnford-Slater's assault flotilla ran into a German convoy which was *en route* from Boulogne to Dieppe. A battle broke out at once and Durnford-Slater's gunboat was so badly hit that nearly half of its crew and Commando passengers were killed or wounded and the craft was wrecked. None of this was Durnford-Slater's fault. The destroyers which were supposed to escort the twenty LCIs had gone off somewhere in the English Channel. Nobody ever explained why the flotilla commander took this disastrous action, which left the lightly armed LCIs terribly vulnerable.

Major Young, with only eighteen men, was with Lieutenant-Commander Buckee, who was as resolute as Young himself. Young described the exchange between them as they neared the enemy shore:

'There you are,' Buckee said. 'That's your beach.'

'What do we do now?' Young asked, all too well aware that No. 3 Commando's original plans had been wrecked.

Buckee replied, 'My orders are to land, even if there is only one boat.'

'Well, those are my orders too,' said Young. 'We are to land whatever happens, even if we have to swim.'

A veteran of Dunkirk, Guernsey, Lofoten and Vaagso, Young hoped that he might still be able to do something useful at Dieppe. Unopposed, he and his men climbed the cliff which confronted them, with the considerable but unintentional help from the Germans – they had placed barbed wire all the way up the cliff face and in doing so hammered in pegs on which the Commandos now found foothold while climbing.

German troops overlooked Young from a church tower and he hoped to climb up and winkle them out but somebody had removed the tower's long ladder. The Commandos then tried to work their way behind the battery which was their principal objective but unseen riflemen kept them pinned down.

Young ran his men into a cornfield and with concealment from the tall stalks they steadily approached the battery until they were only 200 yards from the guns. From here they sniped at the Germans, who were well protected behind concrete walls. Irritated by the British small arms fire, one gun swung around and fired its 6-inch shell at the Commandos. It was noisy and alarming but the gunners could not depress the barrel sufficiently to hit the Commandos. After four rounds the Germans gave up this pointless firing, and it is possible that the Commando sniping had knocked out the crew.

Despite the fire discipline of Young's Commandos they were running short of ammunition and he knew that before long enemy infantry and tanks would turn up to support their gunners. Young shot up an observation post on the cliff and then got his men down to the beach, where Buckee was waiting for them. Young's troop and Lovat's No. 4 Commando had done well but elsewhere chaos reigned and the raid was foundering.

It was a terribly costly failure. The Royal Navy had 550 casualties and lost a destroyer as well as many landing craft. The Army's casualties, mostly Canadian, numbered 3,670. Among the large amounts of *matériel* lost were 29 Churchill tanks. The Germans admitted to a loss of 591 men as well as several guns destroyed. The Royal Air Force lost 106 aircraft and 153 aircrew, another appalling cost of the raid.

Peter Young had a story to tell about Dieppe after the war. Some US Rangers were attached to the Commandos and a few were captured. This was the first time the Germans had taken American prisoners and they were interested to interrogate them. One Ranger, a man at least 6½ feet tall, was asked, 'How many American soldiers are there in England?' Like all Allied soldiers, the Americans had been taught to give no information other than name, rank and serial number but this angry Ranger responded, 'There are 3 million. They are all as big as I am and they have to be kept behind barbed wire to stop them swimming the Channel to get at you bastards!' As Young said, 'Fortunately this particular German officer had a sense of humour.'

As the victor in a battle that had demolished the enemy, he could afford to be amused. Not all the Germans were arrogant Nazis. They had captured the brave but battered Captain Pat Porteous and when the news came through to Offlag VIIB that he had been awarded the Victoria Cross for his magnificent courage and leadership at Dieppe the Offlag's German officers held a parade to celebrate the award. Presumably Adolf Hitler was not asked to approve of this compliment to a British hero – and a Commando at that! Hitler hated the British Commandos.

Lord Lovat, victor of the assault on the Varengeville guns wrote a considered report on Dieppe. This is its essence:

One thing is certain – never postpone an operation and then remount it after an appreciable delay. I also think that it was a mistake to concentrate shipping at Newhaven, immediately opposite the target. It was quite certain that German reconnaissance aircraft never lost sight of this fact, although the German High Command were maybe unaware of the exact target, and it stands to reason that precautions were redoubled on the other side. With respect to the Canadian forces, I do not think that they were sufficiently aware of the magnitude of their task. High courage and ordinary infantry training are no answer to machine-gun fire and impregnable fortifications, set in cliffs. By the same token, I do not think either the planners at Combined Operations, or the tie-up between Army, Navy and Air Force had at that time reached the high standard that was subsequently achieved.

Lovat claimed no particular credit for the Commandos but at Dieppe they demonstrated that Achnacarry Commando Centre had trained them well. The Canadians' sacrifice at Dieppe in no way diminishes their courage. Their bravery was exemplary, their soldierly skill impressive. Even the German victors admitted this.

Mountbatten has been heavily criticised for the disaster of Dieppe but some of his peers did praise him. On 10 June 1944, just four days after D-Day and the Normandy invasion, six of the best known names of the war were together in a train travelling from Portsmouth to London. They had just returned from a visit to the Normandy battlefront and they composed a letter to Mountbatten, who was now Allied Supreme Commander in South-East Asia. They did not mention Dieppe but it is clear from their phraseology that they had this operation in mind:

Today we visited the British and American armies on the soil of France. We sailed through vast fleets of ships with landing-craft of many types pouring men, vehicles and stores ashore. We saw clearly the manoeuvre in progress and in process of rapid development. We have shared our secrets in common and helped each other all we could. We wish to tell you at this moment that we realise that much of this remarkable technique and therefore the success of the venture, has its origins in developments effected by you and your staff of Combined Operations.

The six men who signed this note were: Winston Churchill, Field Marshal Lord Alanbrooke, Chief of the General Staff; the South African Field Marshal Smuts, an adviser to Churchill; General George Marshall, Chief of the US Combined Chiefs of Staff; the US Admiral Ernest King; and General Hap Arnold, one of the two most senior US

Air Force generals. In effect, they were telling Mountbatten that in their opinion Dieppe had been worthwhile. It is difficult for history to concede this but certainly what happened as a result of the fiasco helped the Normandy invasion to be the success that it was. For instance, after Dieppe it was obvious to all the planners and even to Churchill that trying to capture a German-held port would be an exercise in suicidal futility. This is why the Mulberry Harbours came into existence; the invaders took their port with them. This was much cheaper in lives than trying to seize one from the capable brave and stubborn German defenders.

Churchill may have had something on his conscience when he helped to write this letter to Mountbatten and I speculate that it was he who suggested that the letter be written. He certainly should have had an unquiet conscience concerning his role in Dieppe. It was he who insisted on no RAF bombing raid before the assault.

General Montgomery, who was leading the invasion of Normandy when the other notables were penning their letter to Mountbatten, was also blameworthy. It was he who insisted on the suicidal frontal attack across Dieppe's main beaches. Later, he said, 'I believe we could have got the information and experience we needed without losing so many magnificent Canadian soldiers.' From Montgomery, who was never inclined to take the blame for anything that had gone wrong, this observation was tantamount to a confession of incompetence as well as a grudging apology.

A few years after the war, Churchill and Mountbatten each commented on Dieppe. I quote here the most significant parts of their accounts:

Mountbatten: 'Dieppe was one of the most vital operations of the Second World War. It gave to the Allies the priceless secret of victory. If I had the same decision to make I would do as I did before.'

But as the chief instigator of the raid and its overriding planner, he would say this, wouldn't he?

Churchill: 'Dieppe occupies a place of its own in the story of war and the grim casualty figures must not class it as a failure . . . honour to the brave who fell. Their sacrifice was not in vain.'

But as the leader who ultimately had to give approval for the raid and then carry the ultimate responsibility, he would have to say this, wouldn't he? It is unsavoury to see Churchill exploiting the dead servicemen's sacrifice in order to stifle criticism. Mountbatten's defence lies in what he calls 'the priceless secret of victory' – but he does not reveal this secret. This, too, is a form of deception and equally unsavoury.

18

THE SINGAPORE RAIDS 1943–4

JAYWICK AND RIMAU

Following the British, French and American forces' collapse early in 1942 under the speed and sheer force of the Japanese, who controlled all of South-East Asia, it was considered imperative for the Allies to hit back. The intention was laudable but where could any retaliation begin? It had to be a success, for a failure would be handing the Japanese a propaganda victory. Raids were an obvious answer and Singapore and Saigon were among the targets proposed. Singapore was a desirable objective because it was here that the British had been subjected to a humiliating defeat. Several plans were put forward but Allied Intelligence Bureau (AIB) and British and Australian GHQs believed that the risk of failure was too great.

Among the young officers who were eager to raid enemy targets was Captain Ivan Lyon, aged twenty-eight, of the Gordon Highlanders, who had been based in Singapore. Not only was he a tough soldier but an experienced sailor; during leave periods he had made many trips in ocean-going yawls throughout the countless islands south of Singapore. He spoke Malay and, weathered from the hot sun and wearing little clothing, he even looked like a Malay.

When the Japanese invaded Malaya on 8 December 1941 the Special Operations Executive (SOE) in London organised some 'stay-behind' parties, collectively known as Force 136, which would harass enemy lines of communication. Lyon was enthusiastic to be recruited as a guerrilla leader but his superiors ordered him out of Malaya; he was 'too valuable' to risk. Lyon reluctantly obeyed but told his colonel, 'I'll go, but I'll make damn sure to be one of the first back.' He was as good as his word. From the sabotage training school where he was an instructor he took with him a batman, Corporal R.G. 'Taffy' Morris. With them went another valued officer, H.A. 'Jock' Campbell, former manager of a rubber plantation.

In and around Sumatra Lyon and Morris saved many civilian victims of refugee ships which had been bombed and sunk by Japanese planes. For their courageous work and service Lyon was decorated MBE and Morris BEM. During these dangerous days Lyon met Bill Reynolds, a 61-year-old Australian, a veteran of the islands, who like Lyon, had original ideas about hitting the Japanese.

Reynolds owned the *Kofuku Maru*, a yawl which he had brought thousands of kilometres through dangerous waters all the way to India, where he and Campbell met Reynolds again. Lyon, with the same audacity that had made David Stirling the scourge of hidebound senior officers, told British GHQ in India that he proposed to sail the *Kofuku Maru* back to Singapore with Commandos and collapsible canoes. The raiders would paddle into Singapore roadsteads by night and attach limpet mines to Japanese ships. When officers of intermediate seniority scoffed at his ideas Lyon took them direct to General Sir Archibald Wavell, Commander-in-Chief India, and Admiral Sir Geoffrey Layton, Commander-in-Chief East Indies Station. Both these men backed Lyon.

Reynolds, having renamed his ship the *Krait*, after a small but deadly snake, made two attempts to sail it to Australia. Its ancient engine was not up to the voyage and the *Krait* was sent to Sydney as deck cargo on a tramp steamer. Lyon, Campbell and Morris, still planning their forays, went on ahead to report to AIB in Melbourne. Here Lyon learned that a German raider had intercepted the ship which was supposedly taking his wife and son to safety. Then the Germans had transferred the passengers to the Japanese, who had interred them in Tokyo.

Stung by this bad news Lyon's plans were now invested with a personal desire for revenge. However, again like David Stirling, Lyon could not interest his superiors in Melbourne in his raids. General Douglas MacArthur, the American who was now overall commander of Allied forces in the Pacific, curtly dismissed Lyon. The young officer's invocation of Wavell's name as a supporter only further irritated the egotistic MacArthur. But again, like Stirling, Lyon was a determined, polite and skilful lobbyist and he managed to get an introduction to the Governor-General of Australia, Lord Gowrie VC.

Gowrie, a splendid warrior in his day, instantly saw that Lyon should be encouraged. Through him Lyon was put under the authority of Lieutenant-Colonel G.S. Mott, Chief of the Special Reconnaissance Department (SRD). Mott and Lyon, both British, got on well together, but Mott knew that he could not sell Lyon's ideas to MacArthur and GHQ without some bold Commando operation as a precedent.

In the meantime, the Z-Special Unit had been established at Townsville, Queensland, under Major A.E.B. Trappes-Lomax. Among his officers, all thrusters and natural raiders, was Lieutenant Sam Carey of the AIF. Carey was pushing a proposal for an attack on Japanese shipping in Rabaul harbour, New Britain. As liaison officer for Z-Special, Carey had access to Commander-in-Chief New Guinea Force, Lieutenant-General Sir Edmund Herring, and Commander-in-Chief Land Forces, General Sir Thomas Blamey. Z-Special came under the authority of Herring, not MacArthur.

In January 1943 Carey put a detailed plan, Operation Scorpion, to Blamey and Herring. A submarine would take a small group of selected Commandos, with their canoes, to a drop-point 16 kilometres off Rabaul. They would penetrate the harbour, attach limpet mines and then hide on Vulcan Island, which Carey knew well, until Japanese uproar had died down. Blamey was frank: the party would be caught and shot but he gave Carey *carte-blanche* authority to do whatever he wished during his planning. By the end of March 1943 Carey had assembled nine AIF men, Captains R.H.C. Cardew, A.L. Gluth, D. Macnamara, Lieutenants R. Page, J. Grimson and R. Downey, Company Sergeant-Major G. Barnes, Sergeant H. Ford and Corporal G. Mackenzie. After three months of rigorous training Operation Scorpion was ready for raiding.

Ivan Lyon had also been busy discussing his Singapore plans with Colonel Mott. Mott had the idea of testing the effectiveness of well-trained Commandos by mounting a raid, using dummy limpets, on a tightly guarded Allied port. Mott was too shrewd to put anything on paper but he casually mentioned the idea to Sam Carey. The enthusiastic Carey chose Townsville, a busy harbour full of troop transports, merchantmen and naval escort vessels. Because the place was vulnerable to Japanese air and sea attack, security was rigid and it was regarded by the authorities as impregnable. The narrow entrances and approaches were mined and a mine-control point was sited at the extreme end of a long breakwater. If the observers stationed there spotted Carey's canoes they would assume them to be enemy intruders and electrically detonate the mines.

This was to be one of the most hazardous raids of the war – even though no genuine destruction would take place. In five canoes, each crewed by two men, the Scorpion raiders left their base on Magnetic Island and paddled to their objective at midnight on 20 June 1943. Carey and Mackenzie attached limpets to two American Liberty ships, used as troop transporters, and to a Dutch ship; Cardew and Barnes fixed their mines to two destroyers and another ship; Gluth and Page chose a ship named *Akaba* but because it had a barge moored alongside, they had to fix their limpets above the waterline. Downey and Grimson had limpeted two ships and were in the process of dealing with a third when a sailor, smoking as he leaned over the side, asked, 'What are you doing there?' 'Just paddling around', Grimson said casually. Flicking his cigarette butt into the water, the sailor said, 'Good night, mate' and disappeared.

With their limpeting complete, Carey's men met at Ross Creek, dismantled and hid their boats and went to Townsville to sleep. At 10 a.m. the skipper of the *Akaba* reported 'something strange' attached to his ship. Soon other ships contacted Townsville security HQ about 'suspicious objects'. In no time the alarm was raised in GHQ in Melbourne and with the Ministry of Defence in Canberra. One priority signal went to Colonel Mott's office but Mott was carefully 'unavailable'. Colonel Alison Ind, MacArthur's intelligence chief, knew where the blame lay. 'Find Lieutenant Sam Carey', he ordered.

Awakened from a relaxed sleep at 3 a.m., Carey was arrested and paraded before senior naval officers. He explained his mission and produced General Blamey's 'letter of authority'. This did not impress the Navy. Carey offered to remove the mines but the captains of the many ships affected would not hear of this. Convinced that the mines were real, they thought that Carey might accidentally set them off.

Colonel Mott, when finally tracked down, was vague about any authorisation for the raid, but he came down strongly in praise of the efficiency of Carey's team. The Navy released the Army lieutenant but insisted that AIB transfer him to New Guinea for conventional military service. In this way the Allied war effort was deprived of the services of an ingenious and determined raider.

GHQ applied the criticism which was so often aimed at enterprising ideas – Carey's 'raid' had been 'irregular'. Carey had been sacrificed but senior planning officers were by now convinced that a genuine raid against enemy shipping was worth attempting. As Mott had speculated, the example of a successful 'raid' had impressed even MacArthur.

Ivan Lyon and his team had been in training for months and Lyon, delighted by Carey's 'destruction of Townsville harbour', as he called it, made use of the lessons that it produced. But the situation was still not easy for this innovator. To create a small staff, he had to make use of £30,000 which his supporters in India had given him. His key officer in Melbourne was Jock Campbell, a fine administrator and organiser.

Lyon was at last able to schedule his raid. His operation was code-named Jaywick, for no apparent reason, but the name of the target, Singapore, was never mentioned. He selected Lieutenant Donald Davidson RNVR as his second-in-command while the *Krait*'s captain and navigator was Lieutenant H.C. 'Ted' Carse RANVR. Lieutenant Bob Page, who had been one of Carey's team, was taken on as medical officer although not yet fully qualified as a doctor. At twenty-one, Page was already a trained Commando. Corporal Taffy Morris was officially the medical orderly but he was a man of many skills. Leading Stoker J.P. 'Paddy' McDowell, succeeding the brave but ageing Reynolds, would run the *Krait*'s new 105 hp Gardiner diesel engine. The rest of the team were Leading Stoker K.P. 'Cobber' Cain, Leading Telegraphist H.S. Young, and five Able Seamen, W.C. 'Poppa' Falls, A.W. 'Arty' Jones, F.W. 'Biffo' Marsh, M. 'Moss' Berryman and A.W.G. 'Happy' Huston, aged eighteen. Corporal A. 'Jock' Crilley was appointed cook, but he was also a trained Commando.

The men believed that they were headed for Surabaya, Indonesia; the officers knew differently but Lyon did not disclose the true destination until 5 September, three days after leaving Exmouth Gulf, Western Australia. The men took this news quietly, though all knew that Singapore was an infinitely more difficult target than Surabaya. The actual raiders, with Lyon, were Davidson, Page, Falls, Huston and Jones, but the rest of the crew, under Cain and Carse, would have a stressful and dangerous two weeks while they waited for the raiders to return. On 16 September, *Krait* was approaching Temiang Strait and was within 160 kilometres of its target. The pilot of an enemy plane, flying at 'zero feet', took a good look at *Krait* and Lyon and Davidson, the only men on deck, only just had time to put on conical native hats and wave a friendly greeting. Two patrolling seaplanes also showed interest in *Krait* but no apparent suspicion. Lyon noticed much evidence of extensive sea and air patrols, together with watchtowers: it was not going to be possible for *Krait* to lie camouflaged in some deep inlet to await the return of the raiders. Changing plans Lyon told Carse to drop the raiders at Pompong Island, 50 kilometres from Singapore, and then take *Krait* on a run to Borneo before returning for rendezvous in 48 hours.

Lyon made a final check to the Folboats and the team's equipment – limpet mines, revolvers, fighting knives, charts, binoculars, medical supplies and water and food for a month. After handshakes, the six raiders set off for Singapore. Lyon, if nobody else, was well aware that there was not another Allied fighting man within 3,200 kilometres of them. They spent two days on an island toning up their muscles, slack from forty days at sea. Their planned attack-base island was found to be unsuitable, so Lyon chose the tiny island of Subor, due south of Singapore. The allotment of targets was Davidson's responsibility as attack leader. He and Falls would go for shipping in Keppel harbour, Lyon and Huston were given Examination Anchorage and Page with Jones had a free hand at Bukum Island.

On the night of 26 September Lyon issued 'sweets' – actually a cyanide tablet. One bite and five seconds later any man who might be captured would be beyond Japanese vengeance and torture.

The Japanese were so confident of their security that Singapore was one great blaze of light, with water traffic moving freely without escort. At Keppel, Davidson and Falls slipped their Folboat through the boom and limpeted a heavily laden freighter and two other targets, before paddling for Batam Island to find a hiding place. Page and Jones inspected Bukum Island, found a large freighter riding high and attached three limpets, then they fixed limpets to a modern freighter and a heavily laden tramp. Their proposed hiding place was Dongas Island. At Examination Anchorage, Lyon and Huston chanced upon a really good target – a laden oil tanker, which they limpeted close to the engine room and propellers.

Then there was danger. Lyon looked up to see a man staring down at them from a porthole. Like all good raiders, Lyon instantly tried bluff. He whispered to Huston to paddle on steadily as if they were making an inspection. The man kept on looking but after a few minutes he turned away. Lyon and Huston attached their last limpet. Then they too raced to Dongas. Here, Lyon and Huston, Page and Jones hid their canoes and climbed the island's only hill to wait for the explosions. The first was at 5.15 a.m. and within 20 minutes they heard six others. Singapore was blacked out and many sirens sounded. In the light of dawn all six men saw thick smoke pouring from the burning tanker. Lyon, Davidson and Page estimated that they had sunk 50,000 tons of shipping, but it actually amounted to about 35,000 tons.

A superb navigator and smallship skipper, Ted Carse collected the six raiders and set off for home. He knew that the danger period would be when *Krait* was passing through Lombok Strait. He chose to do so at night but at 11.30 p.m. on 11 October a Japanese destroyer of the Sigure class emerged from the bright moonlight at a speed of 25 knots. *Krait* was capable of a mere 7 knots. Carse ordered action stations and the men, Bren machine-guns and Tommy-guns at the ready, hid behind the ship's concealing sides.

Carse had many times rehearsed for this moment in his mind and as the sleek enemy destroyer drew parallel with *Krait* and slowed to the small boat's speed he knew that this was crisis time. Every man on *Krait* had been instructed in the actions that would follow a challenge. If ordered to stop, Carse could try to draw alongside the destroyer so that his men could fight. When that proved hopeless, as it must do, whoever survived would detonate the large quantity of explosives in *Krait*'s bilge, sinking both ships.

Alone on deck and looking like a native, Carse had set a course for Bali, not a suspicious destination. Not even glancing at the destroyer, Carse kept to his course. The destroyer closed to within 100 metres, pacing *Krait* minute by minute. Behind *Krait*'s sides Lyon and the others sweated in apprehension, waiting for a challenge or a burst of fire. For a full half hour the officers on the destroyer watched *Krait*, then it abruptly veered away at high speed, heading for Lombok. The Jaywick warriors were safe and they celebrated with a bottle of rum which Carse produced and with Crilley's 'Singapore dinner', a lavish meal of beef, rice and onions.

Carse anchored *Krait* in Exmouth Gulf on 19 October. In 47 days he had taken the little ship 8,000 kilometres and for 33 days it had been in enemy territory. Without losing a man, Jaywick had hit the Japanese in the heart of their stolen empire. It had been an epic raid. Lyon was called to Melbourne to make his report and to be promoted to major. Carse sailed *Krait* to Darwin and handed it over to the SRD base.

Lyon, Davidson and Page were each awarded the DSO; McDowell, Falls, Jones and Huston received the DSM, Morris and Crilley the MM. Carse, Young, Cain, Marsh and Berryman were mentioned in despatches. Everybody felt that Carse had been slighted with this least impressive recognition and it was known that Lyon himself had recommended him for the DSO or DSC.

The gallantry awards were approved in 1944 but by the time they were announced in August 1946 most of the men were dead. Like so many other raiders, the Jaywick team had to be satisfied with the knowledge that they had carried out a magnificent mission.

The success of Jaywick infuriated the Japanese and put them on their guard against further raids on Singapore. Despite the increased dangers, Lyon, now a lieutenant-colonel, prepared for a second Singapore raid, Operation Rimau. He ignored the misgivings of his superiors and by some of his Jaywick comrades, who thought he was pushing his luck too far. Lyon recruited and trained twenty-three operatives, including six from Jaywick, and training began early in 1944. Lyon chose the new Motor Submersible Canoe or 'Sleeping Beauty', instead of Folboats he had used for Jaywick. The SB's handicap of being too slow to fight the current flowing south from Singapore was never properly considered. With fifteen SBs and eleven Folboats, the Rimau team left Western Australia on board the British submarine HMS *Porpoise* on 11 November 1944.

Porpoise left stores on Merepas Island, Rimau's rear base, and went on to Pejantan Island, the forward base. On 28 November a boarding party seized the Malayan junk *Mustika* to use as a mother ship and carry the raiders and their gear close to the target. *Porpoise*, with *Mustika*'s native crew aboard, returned to Australia. A sister sub, HMS *Tantalus*, would pick up the raiders at Merepas on 8 November. Should *Tantalus* fail to arrive, Lyon and his men were to wait for 30 days and if no rescue vessel had arrived by then they were to make their own way home.

This was an irresponsible plan. Were the operation to be successful – and even it were not – the area would be combed by entire divisions of Japanese soldiers and many patrol boats. The men could not possibly hide for thirty days.

On 1 October *Mustika* returned to Merepas to leave a party of men to guard the stores. Lyon was now ready to begin his raid and *Mustika* anchored off Laban Island, only 16 kilometres from Singapore's Keppel harbour. Lyon waited for darkness to launch the SBs but a crisis now occurred. A Malayan Water Police patrol boat approached *Mustika* to make a routine inspection. Somebody on board *Mustika* opened fire instead of waiting for the three Malays to come aboard, where they could have been quietly dealt with. Two policemen were killed but the third swam ashore and raised the alarm.

With the mission's secrecy destroyed, Lyon and his naval co-leader, Lieutenant-Commander Davidson, aborted the mission. The SBs were smashed and dropped overboard and when night fell the Folboats were launched and the *Mustika* was sunk by hull charges. In four parties, the Rimau men paddled for Merepas – with no prospect of rescue for the next five weeks.

The Japanese High Command in Singapore sent Captain Tomita and a company of troops to search for the intruders. Tomita was killed and this in itself was enough to rouse the Japanese to full and angry alert. Major Toshida with a complete battalion was sent to search every island in the area. Details of the hunt are sketchy but the basic elements are known. All four escape parties met on Sole Island where they fought Toshida's troops. On 15 October Lyon ordered the rest of his men to run for it while he and Lieutenant Ross covered their withdrawal. Lyon and Ross were shot to death in this heroic stand. Lieutenant-Commander Davidson and Lieutenant Campbell, badly wounded on Tapei Island, committed suicide with their last bullets. Warrant Officer Willersdorf, Lance-Corporal Page and Private Warne paddled to Kalapanaga Island where Warne, seriously ill with malaria was captured and killed. Canoeing a long distance to Romang Island, near Timor, Willersdorf and Page hid out until natives betrayed them to the Japanese in March 1945. They died in prison on the island. Captain Bob Page took refuge in a village but was captured.

Lieutenant Serjent, Lieutenant Reymond and Corporal Craft paddled by night and hid by day in an epic two-month flight eastwards via Sumatra, Java, Bali, Lombok, Flores, Alor and Wetar. Finally they reached the Romang Islands, only 650 miles from Darwin and safety. Their Folboat was wrecked near Maja Island and Serjent pleaded with the crew of a Chinese junk for a passage to safety. Instead, they were thrown overboard. Reymond and Craft drowned but Sergent clung to a log for 10 hours until caught in native fishing nets at Cape Satai, where he was handed over to the Japanese. In all of Australia's war history Lieutenant Serjent's escape attempt was one of the most courageous and determined. He deserved to win through.

In all, eleven members of Rimau, including Able Seaman Marsh who died of illness in captivity, were captured at various places. Taken to Singapore in December 1944 they were kept by the dreaded military police, the Kempei Tai, in Tanjong Pagar prison until February 1945. After exhaustive interrogation about their mission they were charged with espionage and murder. A court martial found them guilty of both charges and they were imprisoned in Outram Road gaol, Singapore. The court martial was a mere formality and in terms of international law the verdict was a farce. The Rimau men had been in uniform so 'espionage' was an irrelevance. As soldiers they had fought an armed enemy in time of war and could not be guilty of murder.

On 7 July 1945 all ten men were ceremonially beheaded, a death claimed by the Japanese to be in respectful recognition of the Rimau team's heroism. According to reports, a week after the executions General Igitaki, Commander-in-Chief Singapore, addressed the full staff of the 7th Area Army on the subject of Operation Rimau. He said: 'We Japanese have reason to have been proud of our bravery and courage in action, but these heroes showed us a fine example of what true bravery should be. Unless we try much harder to make ourselves better soldiers, we ought to feel ashamed of ourselves before these heroes.' Perhaps it was obvious to Igitaki that the war was ending and that the Allied victors might want to punish him for killing the Rimau men – and others. If so, his words were just a form of insurance; he could plead that he was not guilty of wanton murder but that he had honourably dealt with his captives.

Heroes the Rimau men undoubtedly were – and of a special kind.

They ventured far into enemy territory without hope of rescue or assistance of any kind should something go wrong. Rimau was a lesson for planners of raids. The arrangements for making rendezvous with the raiders on their departure from Singapore were grossly inadequate and negligent. That somebody aboard the *Mustika* shot at the Malay policemen indicated lack of discipline and control. It also showed poor marksmanship since one of the policemen lived to report the raiders' presence. Heroism is never enough. There must always be a reasonable chance of survival for men who are not suicide attackers. Very few of the Second World War's raiders were hell-bent on suicide; the Japanese kamikaze pilots were the extreme example.

After the unqualified success of Jaywick, Rimau was an unmitigated disaster. It was a sad lesson on the penalty for failure in a raid. Several men were to blame in one way or another. Superior officers should never have permitted Rimau to go ahead in the first place: to make a raid a second time at the same place was foolish, especially as there were other good targets, such as Rabaul. The conducting officer, a captain, was negligent in arranging the escape and pick-up plan: there had been indications that Lyon and Davidson, still aflush with the triumph of Jaywick, had become too gung-ho for responsible leadership. Also, the raid lacked a 'ship's captain', as Carse had been for Jaywick: it is unlikely that Carse would have agreed to the idea of seizing a native boat about whose performance he could know nothing. Rimau was from the beginning an adventurous indulgence.

19

AUSTRALIAN RAIDERS
PAR EXCELLENCE

Australian Commandos and Independent Companies made many raids against the Japanese in the great islands north and north-west of Australia during the years 1942–5. Two men stand out as repeatedly successful raiders in one of the most hostile environments of the war and against a formidable and ferocious enemy. They were Major Rex Blow and Captain Robert Kerr 'Jock' McLaren. Both had been members of the Australian 8th Division which had been captured by the Japanese during the disastrous Malaya–Singapore campaign of early 1942. Among the few soldiers to escape from Japanese captivity, Blow and McLaren and a few other Australians were anxious to return home and fight with units of the AIF but at Mindanao the senior American officers in command of guerrilla operations had other plans for them.

To begin with, the Australians were put to work as coast watchers on Tawi Tawi Island, with Rex Blow as their leader. Nobody could then have foreseen that he was to become the most outstanding guerrilla leader of the southern Philippines. As coast watchers, the Australians spotted enemy shipping and flashed radio reports to US Navy ships. They faced danger not only from the Japanese but from the fierce Moro tribesmen, large numbers of whom were pro-Japanese because Japanese propaganda had turned them against the Americans and their allies. Blow heard that the Moros intended to capture Tawi Tawi and decided on a pre-emptive strike. With McLaren, two other Australians and three native fighters, he set off to attack the Moros in the immense old Spanish fort which they had made their base. But the Moros ambushed the Australians as they were crossing a creek, killing one of them and wounding another. With his men vastly outnumbered, Blow picked out the Moro leader and killed him at close range with a burst of Tommy-gun bullets. Never having experienced the lethal fire of a machine-gun, the Moros were demoralised and fled into the jungle.

Ordered by radio to leave Tawi Tawi and go to Mindanao, the

Australians made a long and dangerous journey to Laingan, Mindanao, which the Japanese attacked in strength on the night of 21 December 1943. In the fighting that followed McLaren led a patrol to scout Japanese movements. Slipping into the jungle, he soon came upon enemy in heavily defended positions. A born raider, McLaren attacked and fired two full magazines at the Japanese until his gun jammed and he had to run. Blow led four other Australians to a ridge from where they could see Japanese unloading stores from boats. Holding their fire until the number of working Japanese soldiers had increased to about 100, the Australians opened up with devastating effect, causing many casualties.

A submarine arrived from Australia late in December and for intelligence purposes took out two of the Australians. The submarine also brought notice of McLaren's promotion from sergeant to lieutenant while Blow was promoted to major, skipping the rank of captain. The 27-year-old Blow had impressed the American–Filipino guerrilla command and he was appointed Chief-of-Staff of the 108th Division, Tenth Military District. It was a position of high responsibility and while Blow would have preferred to return home to Australia he liked the idea of organising his own raids.

McLaren's new assignment was to transform about 500 Filipino peasants into soldiers. He disliked large numbers of men as a fighting force, especially as raiders, so he selected twenty whom he regarded as reliable. Under Blow's command, the small group of raiders were so effective in maintaining secrecy that the enemy withdrew after finding only one of the guerrillas' many limited-range radio transmitters.

In mid-March 1944 the Australians heard that the Japanese were again intending to campaign in the southern Philippines and that Tawi Tawi was a priority for capture. Blow and McLaren positioned their twenty trained Filipinos on a low hill covering the most likely enemy landing beach. Japanese troop-carrying barges unloaded their soldiers, who milled about on the sand. On Blow's signal the raiders opened up with machine-guns and raked the Japanese, practically wiping them out. As other barges began to bring in reinforcement troops, and destroyers opened fire on their hill, Blow's raiders withdrew.

Establishing a temporary base in the jungle, Blow and McLaren decided to 'celebrate' Emperor Hirohito's birthday with the Japanese. They knew that the enemy would be on parade in large numbers and observation told them that the ceremony would take place on grassy fields near the beach. McLaren began the Australians' contribution to the party by firing mortar shells into the massed ranks of soldiers. Soon the grass was strewn with dead and dying and the survivors ran into the sea. The hidden raiders cut down many with machine-gun and rifle fire. Again it was time to withdraw before reinforcements arrived.

After several raids on enemy bases, the great strain and effort, as well as jungle diseases had reduced McLaren to about 7½ stone. He was ordered to rest but he could not allow Rex Blow to take all the responsibility and he kept going. In one way he was living even more dangerously than Blow because the Japanese now had his name and photograph. In Changi prison, Singapore, they had deprived prisoners of their army paybooks, which carried the soldier's photo and now they enlarged McLaren's image and posted it at public places in the Philippines. He was worth 70,000 pesos, dead or alive, a price which McLaren regarded as 'reasonable'.

He had a realistic view about the Japanese as soldiers and he often stated it for the benefit of the Filipino guerrillas and his own mates. He said, 'You have to remember that when you bump into a Jap on the track he gets just as a big a shock as you do. You stand there like gawks for a couple of seconds, then the man who is first on the draw wins the box of chocolates. Don't believe that the Japs are super-soldiers. Their bowels move just as quickly as anyone else's.'

But the enemy was always only one of the dangers these soldiers faced. All men trained as raiders in Australia were taught how to remove their own appendix should no expert help be available. Unfortunately, however, Jock McLaren was without the benefit of this training; he had gone straight from a POW camp into his own type of war. The day came when he knew that his appendix had burst. If it were not removed he would die, but Rex Blow was operating elsewhere and McLaren was alone except for a friendly Moro chief, his Filipino wife and a young native. Furthermore, he had no anaesthetic. In civil life he had been a veterinary surgeon so he had some knowledge of organs and now he instructed his native friends how to prepare for the operation. They set up floorboards on boxes as an operating table and kept a pot of water boiling. From the Filipino woman, McLaren obtained two large spoons, which the young native bent as retractors. McLaren asked for them to be sterilised, together with a pair of scissors, a needle and a razor blade. The woman found a pair of forceps, which were also sterilised. It was she who turned banana leaf into sutures. Finally, McLaren asked his friends to lift him onto the table and hold a mirror over his abdomen. With the razor blade he cut through skin and tissue and prised the muscles apart. The Filipino woman held them open with the bent spoons. The appendix had ruptured, as

McLaren had expected, and it was adhering to his intestines. Cutting it away, he pulled it through the incision. The woman instinctively knew to swab the blood and fluid.

Trembling and in pain, McLaren worked the displaced muscles back into position. Asking for the sewing needle and banana thread, he forced himself to stitch the wound. Every motion was agony. There had been many pauses as he regathered his strength and the operation lasted four and a half hours. Then McLaren slept. He hoped for enough time to recuperate but after only three days a messenger arrived to warn him that the Japanese were approaching. McLaren limped into the jungle. Sheltered by native friends in one safe house after another he was fighting fit within two months. Few raiders of any nationality have shown such toughness and will.

He heard that he had been promoted to captain but much better than this well-deserved promotion was the arrival from Australia of a 32 hp whaleboat, lashed to the hull of a submarine. Blow had asked for it several months earlier and now he put McLaren in charge of it as a seaborne raider. Ecstatic with this gift, McLaren christened it *The Bastard*. Eight metres long, *The Bastard* was armed with a 20 mm quickfire cannon in the bows, two .30-inch machine-guns amidships and a .50-inch gun aft. After enlisting and training a crew of eight, Blow gave McLaren his assignment: to patrol the islands between Mindanao and North Borneo and report on enemy activity.

Blow had been busy in the preceding months. In November 1944 he led 200 guerrillas in a raid on an important enemy airfield at Malabang, on the coast of Moro Gulf. He carried out his plan of destruction before the Japanese could bring in reinforcements. But he was not finished with Malabang and as a first step he ordered McLaren with *The Bastard* to prevent supplies from reaching Malabang and Parang. It was a dangerous mission but it appealed to McLaren infinitely more than reconnaissance patrolling and to prepare for it he dressed his crew in captured Japanese uniforms – of which the two Australians had a large stock – and hoisted the enemy flag over *The Bastard*.

His most spectacular raid took place in March 1945 when one morning he casually motored his gunboat into Parang. His crew waved a friendly greeting and excited and bored Japanese soldiers rushed to the jetty in welcome. When only 100 metres away McLaren, who always said that he was a stickler for international law, struck the Japanese flag. Lamenting his lack of an Australian flag, he ran up the Stars and Stripes. From his place at the bow gun he opened fire simultaneously with his crew. The fusillade was devastating, killing or wounding all of the assembled enemy troops and sinking three vessels at the jetty. Finally the gunners shot up shore installations. As the garrison began to return fire *The Bastard* sped away, its stern making only a small target.

Neither McLaren nor Blow had ever heard of the exploits of the SBS in the Mediterranean but with *The Bastard* they were achieving exactly the same as the British raiders. McLaren was so satisfied with his naval raiding that he fitted *The Bastard* with a more powerful gun as well as a rocket-firing bazooka, normally used by soldiers against tanks. McLaren used it to sink enemy vessels. The craft and its skipper became the scourge of the Moro Gulf and the nearby seas.

Under orders, in March 1945 Blow laid siege to Malabang airfield, though personally he disliked the idea as unsuitable for raiders. McLaren joined Blow in the trenches. The Japanese knew of his presence and an enemy soldier shouted, 'Come on, McLaren, come out and fight! Why don't you show yourself, McLaren!' McLaren was too wily to shout back and draw fire but he wrote a message, wrapped the paper around a stone and threw it into the enemy lines. He had written, 'You misguided bastards, defenders of Malabang. I will meet you in the open on the beach at Bouchana River. Bring all your garrison and don't forget your aeroplane. After that's finished I will proceed to Tokyo and pull your emperor's nose. McLaren.'

Next day he was aboard *The Bastard* on the Bouchana River when, to his surprise, the Japanese took up his offer to fight. The enemy fighter plane, the only one at Malabang, swooped on *The Bastard* only to be shot down in flames by the boat's concentrated firepower. There followed a machine-gun attack from the shore but *The Bastard* evaded the fire and raked suspected enemy positions. Now with no threat from the air, Blow took the initiative and with McLaren's support from the sea, launched an infantry attack. On 12 April he captured the airfield, drove the Japanese from Malabang and forced them from the Parang area. Blow's action made it possible for the US 24th Infantry Division to land without opposition on 17 April. The Division had expected to take heavy casualties and its leaders were amazed at being able to walk ashore.

Blow and McLaren were used for several more exhausting and dangerous missions but their war ended on 9 August 1945 when the atom bomb was dropped on Hiroshima. Jock McLaren was twice awarded the Military Cross, a rare double decoration in the AIF, which was parsimonious in the issue of awards. The citation, made public on 16 November 1945, was among the longest ever written for an MC award, detailing his extraordinary feats. Rex Blow was awarded the Distinguished Service Order. The Americans wished to bestow decorations on Blow and McLaren but the Australian Army prevented their acceptance. The unstated reason was that senior officers did not want men junior to them to wear foreign decorations they themselves did not possess.

Raiders from Australia's Z-Special unit, singly and in teams, carried out many astonishing raids in Borneo and other places but even collectively they did not achieve the international publicity of similar exploits by their British counterparts in the SAS, SBS, LRDG and other units. Another point must be made about Blow and McLaren. Unlike the British raiders, they had no breaks away from active service. The British had spells of leave in Cairo, Beirut and elsewhere and for a while lived in comfort and without danger. Blow and McLaren were in enemy territory for more than two and a half years. These two Australians were raiders who never rested.

Jock McLaren was killed in an accident in New Guinea soon after the war. In his late seventies Rex Blow was happily growing nuts for a living along the Murray River in Victoria.

20

ITALIAN UNDERWATER RAIDERS, 1941–5

Among British troops in North Africa it was customary to sneer at the Italians as comic opera soldiers, warriors who would rather sing than fight. The hordes of Italians taken prisoner after some of the North African battles, particularly by the Australians, prompted derisive laughter and scorn. And they were not eager to escape, even when this would have been easy. A few score Australian soldiers guarded tens of thousands of Italian prisoners being marched into captivity. 'The Ities [pronounced Eyeties as in pies] couldn't punch their way out of a paper-bag', the British soldiers scoffed. When it was revealed that Italian officers in the desert had stocks of perfume in their tents and dugouts the laughter became more raucous.

The Italian Army had no real raiders in the British sense. Ordinary infantry patrolling could not be called raiding. The only regiment for which the Allied troops came to have any respect was the Bersaglieri, whose men were more dashing and brave.

Unknown to the British, at the time of Italian defeats in the Western Desert, some Italians were under training for raiding of a very special kind. They were the officers and men of the Decima Mas, an élite unit specialising in raids by human torpedoes and assault frogmen. These raiders were eminently successful, claiming more in terms of tonnage sunk than their British counterparts. Decima Mas sank or damaged four warships and twenty-seven merchant ships – a total of about 260,000 tons. Furthermore, they changed the balance of naval power in the Mediterranean.

The Italian Navy had a proud history in the use of manned torpedoes, small torpedo boats and what they called 'special attack' craft. The Grillo had attacked the Austrian fleet's Adriatic base in May 1918 and while it could hardly be called a success it inspired several marine engineers to work along similar lines. On 31 October 1918 Commander Rossetti and his companion Lieutenant

Paolucci took their Mignatti manned torpedo into Pola harbour, Adriatic, and fixed one of their charges to the bottom of the Austrian warship, the 20,000-ton *Viribus Unitis*. Running out of power to get out of the harbour, they primed the fuse on their 'craft' and let it loose in the current. Discovered by the side of the *Viribus Unitis*, the two Italians were taken on board where they learned that their attack had not been necessary. The Austrian Empire had collapsed and its fleet now belonged to Yugoslavia. The Italians at once raised the alarm about their own mines but the warning came too late. The charge blew up, ripping the bottom out of the warship, which sank. Meanwhile the abandoned torpedo, spiralling wildly in the current, collided with the Austrian armed merchantman *Wien*, 7,400 tons, and it, too, sank.

The Decima Mas inherited this raiding tradition. It came under the command of Captain Moccagatta in March 1941 and it was he who developed it into a formidable weapon. For instance, he divided his unit into two parts, one to deal with human torpedoes and assault frogmen, the other with explosive speedboats. Such weapons were essential for the Italians in the Mediterranean because they did not have the weight of warships possessed by the Royal Navy or the French, before France capitulated to the Germans. The Italians were possessive about the Mediterranean, which they traditionally regarded as their lake. The notion that it was a British lake – British admirals and many politicians regarded it as such – was galling.

When Italy joined the war on Germany's side on 10 June 1940, the two-man torpedo, the Maiale (the crews called it 'the pig') was still in its development stages but work now proceeded rapidly. The Maiale was no makeshift craft. It was 7.3 metres long, powered by 60-volt cells and capable of 4.5 knots. The warhead weighed 300 kg. The driver sat astride the craft while his assistant was behind him in a form of cockpit. Various models had different measurements but all were remarkably stable.

Many Maiales were built and the first operation using them was intended to take place in Alexandria harbour, Egypt, on the night of 25/6 August 1940. The plan was for a submarine to clamp four pigs to its casing and proceed at a depth of 30 metres into the enemy harbour, there to release the pigs. This was a risky action because patrolling aircraft could see much further into the water than this, but the Maiales could not safely be taken deeper. However, the submarine detailed for the mission was in training when torpedo planes from HMS *Eagle* caught it on the surface and sank it. The operation had to be abandoned.

The next development was that of mother submarines to transport pigs to their targets. Clamps had given way to steel containers from which the operators extracted them while the mother sub remained submerged.

The Decima Mas crews had many adventures and many mishaps but no successes, though one human torpedo came close to blowing up the battleship HMS *Barham* at Gibraltar. With great courage and determination the Decima Mas crew cleared the harbour boom and were sliding the pig slowly along the harbour bed when its motor stopped. Then Petty Officer Paccagnini's breathing apparatus failed and he was forced to surface. Lieutenant Brinidelli tried for 30 minutes to manhandle the warhead along the sea bed but under the tremendous exertion his breathing apparatus also failed. He set the time fuse but the charge exploded harmlessly and the two Italians were taken prisoner. This close call resulted in more stringent security at Gibraltar.

Three pigs made another attack on Gibraltar in May 1941 but mostly because of mechanical failures they achieved nothing. A few weeks later Major Tessi took a Maiale on a deliberate suicide attack in the Grand Harbour at Valetta, Malta, but without significant success. It has been seriously suggested that Tessi, a brave man deeply involved in Decima Mas, was so frustrated with its inability to strike hard that he sought success with suicide.

With so many warnings of Italian human torpedo activity, the British intensified their harbour defences. On the night of 19/20 September 1941 Lieutenant Catalano and Petty Officer Giannoni reached the entrance to Algeciras harbour, Gibraltar, but while they evaded capture they were forced out into the Straits. They decided now on a target of opportunity, which happened to be the armed merchantman *Durham* (11,000 tons). They fixed their charges with magnetic clamps, scuttled their pig and swam to safety in friendly Spain. *Durham* was so badly damaged that it had to be beached. Meanwhile the crew of a second pig sank the tanker *Fiona Shell* (2,444 tons).

Lieutenant Visinti and Petty Officer Magro crewed the third pig that night. Determined and skilful, Visinti audaciously needled his pig in between the steel cables which supported the boom across the mouth of Gibraltar's military harbour, seeking the carrier HMS *Ark Royal*. He did not locate it but he surfaced close to a British cruiser. He and Magro decided, however, to go for a big tanker; they hoped that when their charge went up the tanker would catch fire and create a hurricane of flames throughout the harbour. They sank the tanker, the Royal Fleet Auxiliary *Denbydale* – which was a major victory, but the fires were contained and did no further damage. Visinti and Magro heard about their success when they ran their pig to Spain, to be applauded by the colony of Italian agents there.

Decima Mas was about to achieve its greatest victory. On the night of 18/19 December 1941, at Alexandria, three pigs launched by their mother submarine, easily managed to enter the harbour by closely following British destroyers returning to port through the boom. That the Italians evaded detection was amazing because British Intelligence had warned of an impending attack and Admiral A.B. Cunningham, the Royal Navy chief, had already ordered urgent and improved security. The target assigned to Lieutenant de la Penne and Petty Officer Bianchi was HMS *Valiant*, a giant of a battleship at nearly 31,000 tons. The run-in was nearly a disaster because Bianchi's breathing gear became choked and he surfaced. The pig now fouled its propellers but de la Penne, a powerful man and a strong swimmer, forced it along the bottom and under the battleship, where he set the warhead's fuse.

The two Italians were captured and when they would answer no questions they were locked in the *Valiant*'s hold, where they would be among the first victims of any explosion. They knew very well that they were close to the warhead but still they did not call the guard and ask to be taken to the captain. The explosion occurred at 6.20 next morning. The *Valiant* was so badly damaged below the water line that it was later used only as a depot ship. With remarkable good luck, its Italian attackers escaped injury.

Meanwhile, Captain Marceglia and Petty Officer Schergat were fixing their warhead under HMS *Queen Elizabeth*, Cunningham's flagship. After the capture of Bianchi and de la Penne the general alarm had been raised but there was not enough time to save *Queen Elizabeth*. The huge explosion briefly lifted the great ship out of the water and it threw Admiral Cunningham five feet into the air, as he later testified. The third Alexandria raid pig, crewed by Captain Martellotta and Petty Officer Marino, had been given a different type of target, the laden tanker *Segona*. This ship was badly damaged, along with the destroyer HMS *Jervis*, which had been lying alongside *Segona* during refuelling.

Decima Mas could hardly have brought off a more important coup, one of the most remarkable of the war. That Hitler and Mussolini knew about it is hardly to be doubted, given the widespread Italian intelligence network in North Africa, with its thousands of Arab informers. Martellotta and Marino had not been caught and were at large and hidden by Italians in Alexandria for three days, long enough to pass their astonishing information to Axis spies. Cunningham took no chances and ordered that all six prisoners be held incommunicado for 6 months. In this way he hoped to keep the raid secret.

The attacks on the two great battleships were a tremendous blow to the British, psychologically and in Mediterranean power projection. But the British have always been masters of military cover-up and Cunningham's officers were frantically busy during the hours after the Italian raid. Somehow, they had to pretend that nothing had happened.

They were helped by the water in Alexandria harbour, which was too shallow for the ships to sink and though sitting on the bottom they remained upright. Officers and men observed the usual naval ritual of Colours. The national anthem was played by the bands, the White Ensign was hoisted and Cunningham took the salute. However, he knew that 5,300 square feet of *Queen Elizabeth*'s hull was stove in.

Orders were given to Alexandria's air defenders that on this morning they should not be over-vigilant in keeping enemy reconnaissance planes away from the harbour. Cunningham and his intelligence officers wanted the Italians to take photographs in the hope that they would believe that the human torpedo attack had failed. But the Italian naval command did learn of the great battleship's helplessness and sent pigs to finish it off. This attack failed. Patched up and refloated, *Queen Elizabeth* limped off to the United States for permanent repairs, which lasted until June 1943.

The raid reversed the strategic position in the Mediterranean for three years. For the first time – and last – in the course of the war the Italian Navy achieved overwhelming superiority and dominated the Mediterranean. With impunity, it resumed supplies to the Italian armies in North Africa and carried out transport missions for the German Afrika Korps in Libya. A few months later the consequent re-strengthening of the German–Italian forces caused the defeat of the British Army and its expulsion from Cyrenaica.

Admiral Cunningham was a fair-minded naval man and later he praised the 'cold-blooded bravery and enterprise of these Italians'. He admitted that the damage to his battleships had been a disaster. The pig swimmers were among the most successful raiders of the Second World War, if only in terms of damage caused. If the Germans and Italians had adequately co-operated to exploit the Decima Mas triumphs they could have altered the course of the hostilities. Rarely have a few raiders – in this case only six – succeeded so majestically. And all survived.

In 1944, after de la Penne and Bianchi returned to Italy from a British POW camp, they were awarded the Italian Gold Medal for gallantry in war. By this time Italy was one of the anti-German Allies. Who pinned the splendid medals on the heroes' chests? None other than Admiral K.W. Morgan, formerly commanding officer of the *Valiant* and now chief of the Allied Naval Mission to Italy. Morgan made a wry comment to de la Penne: 'This is ridiculous – but all war is ridiculous. Congratulations on destroying my ship.' Lieutenant de la Penne kissed Morgan on the cheek.

The Italians had other underwater raiders who were possibly even more courageous than the pig crews. They were the assault swimmers or frogmen, known as the Gamma Group. Formed under the auspices of Prince Junio Borghese when he commanded Decima Mas, the Gamma men caused great problems to the British, especially at Gibraltar and in its Straits.

A Gamma swimmer wore a rubber suit and swim-fins and his breathing apparatus allowed him to stay underwater for up to 40 minutes. On his waist-belt he carried four time-fused charges weighing 4.5 lb and later frogmen were equipped with magnetic limpets weighing nearly 10 lb. A trained Gamma man could operate at a range of nearly 5 miles from his mother submarine. Many Gamma men were taken to their targets in the same subs that carried the pigs.

Unknown to the British, the Italians established a Gamma base under their noses in Algeciras Bay, directly opposite the British naval harbour. It was cleverly done. In 1940 the Italian freighter *Olterra* had been sabotaged by the British and it lay half-submerged in Spanish water. The Italian government went through the charade of handing the ship over to Spain and it was towed into Algeciras harbour. The Spanish were supposed to repair the ship but it was the Italians who did the 'work' on the vessel: they cut an underwater door and through this passed swimmers and occasionally pig crews. By night, with the connivance of the 'neutral' Spanish, men and supplies were slipped into this extraordinary depot ship. It seems likely that the British never learned what was going on.

Beginning in September 1942, Gamma men based in the *Olterra* sank or damaged eleven Allied merchant vessels, a total of 54,200 tons, before the venture ended in August 1943. Also during that period, on 12 December 1942 the submarine mother *Ambra* launched three pigs and ten frogmen into the waters outside Algiers. Their haul was impressive – four merchantmen totalling 22,300 tons.

The British found it difficult to deal with these silent and invisible raiders. They were so alarmed that they formed an anti-sabotage frogmen unit under the naval Lieutenant L.K. 'Buster' Crabb. Learning of this development, the Italians built a booby trap into each of their limpets.

Some frogmen on both sides went 'missing' in the North African waters but the Gamma and pigs operations were remarkably cheap for the Italians. Indirectly, the British benefited from the Italian experience. A pig was washed ashore in Spain, then others were recovered intact at Alexandria and Gibraltar. After close study of these specimens the British produced the Chariot. Later still, after Italy's collapse, the British and Italian underwater teams collaborated against the Germans, the former allies of the Italians. Pigs and Chariots in close alliance attacked Italian ships that the Germans had seized. One of their greatest raiding coups was the attack against the uncompleted aircraft carrier *Aquila* (27,000 tons). The Germans intended to use this large vessel as a blockship at Genoa but a team of British and Italian frogmen sank it in its dock in April 1945.

We return to the essential paradox about the Italians of the Second World War. The Army was spectacularly incapable of producing raiders but the Navy had no shortage of volunteers, nearly all of whom showed aptitude and great skill and, even more importantly, they were men of spirit and daring. The historian primarily responsible for analysing the success of the Decima Mas in comparison with the Army's inability to produce a counterpart on land, speculates that it had something to do with the Latin temperament. He wrote:

The Italians lacked nothing when it came to displaying individual courage and were glad and willing to volunteer for hazardous duty where a man's individual prowess would stand out. They were not, on the other hand, so happy about playing a small part in a large organisation.

Count Julio Valerio Borghese, who commanded the submarine *Scire* which transported the human torpedoes to their targets, after the war described the thoroughness of preparation for the Decima Mas raids:

The operation against Alexandria was most carefully thought out. The most important requirement was the maintenance of absolute secrecy, that indispensable coefficient of success in any

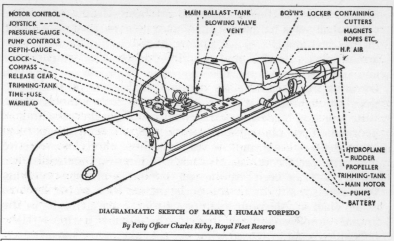

MOTOR CONTROL
JOYSTICK
PRESSURE-GAUGE
PUMP CONTROLS
DEPTH-GAUGE
CLOCK
COMPASS
RELEASE GEAR
TRIMMING-TANK
TIME-FUSE
WARHEAD

MAIN BALLAST-TANK
BLOWING VALVE
VENT

BOS'N'S LOCKER CONTAINING
CUTTERS
MAGNETS
ROPES ETC.
H.P. AIR

HYDROPLANE
RUDDER
PROPELLER
TRIMMING-TANK
MAIN MOTOR
PUMPS
BATTERY

DIAGRAMMATIC SKETCH OF MARK I HUMAN TORPEDO

By Petty Officer Charles Kirby, Royal Fleet Reserve

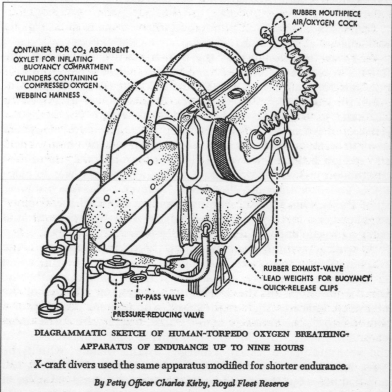

RUBBER MOUTHPIECE
AIR/OXYGEN COCK

CONTAINER FOR CO₂ ABSORBENT
OXYLET FOR INFLATING BUOYANCY COMPARTMENT
CYLINDERS CONTAINING COMPRESSED OXYGEN
WEBBING HARNESS

RUBBER EXHAUST-VALVE
LEAD WEIGHTS FOR BUOYANCY
QUICK-RELEASE CLIPS

BY-PASS VALVE
PRESSURE-REDUCING VALVE

DIAGRAMMATIC SKETCH OF HUMAN-TORPEDO OXYGEN BREATHING-
APPARATUS OF ENDURANCE UP TO NINE HOURS

X-craft divers used the same apparatus modified for shorter endurance.

By Petty Officer Charles Kirby, Royal Fleet Reserve

action. This was particularly so in those actions where the vulnerability of a handful of half-naked men, plunged underwater in the dark depths of an enemy harbour, had to overcome armour-plates, barriers and a hundred methods of watching for and spotting them, and also thousands of people on dry land, operating from behind cover and behind defences on moles and ships, whose business it was to discover and destroy the assailant. Wide use was made of air reconnaissance for the obtaining of information and photographs with a view to keeping us informed about the usual moorings of vessels and the nature of protective measures. Great care was also taken in preparing materials. The senior pilots had been given a long training by myself in carrying out exercises similar to the performance they would have to accomplish in Alexandria. Practice took place at night in the actual conditions prevailing in the enemy harbour, their difficulties being increased wherever possible.

Borghese's emphasis on planning and preparation could equally well apply to all raids undertaken by any nation's forces.

As I have mentioned that the British learned much from the Italians in regard to human torpedoes and midget submarines, it is only fair to disclose that the Germans learned from the British. After the successful British attack on the powerful *Tirpitz* in Kaafiord, Norway, in September 1943, the Germans discovered the remains of X6 and X7 from the deep fjord. The naval designers went to work at once to create Germany's own two-man midget sub, which became known as Hecht, meaning pike. Its official designation was XXVIIA and its fundamental purpose was to place mines under the hulls of Allied ships. However, it was much smaller than the British X-craft. Admiral Donitz, the Nazi naval chief, stipulated, against the designers' advice, that the Hecht should carry a torpedo and its successor carried two.

As with everything else concerning the German war effort, the two-man sub or human torpedo had to be approved by Hitler and in January 1944 he gave his assent. It is not certain whether he saw the prototype but the order, on 28 March, for fifty-two of the new craft indicates high enthusiasm – or perhaps desperation – somewhere in the ranks of senior Nazis. There were to be many variations of the two-man torpedo, with technical differences, but the Seehund (Type XXVII) emerged as the weapon which the Navy would use. At normal speed it had a service range of 270 nautical miles but this could be increased to 500 nautical miles at slow speed.

Everybody involved, except perhaps some of the apprehensive crews, had high hopes for the Seehund when seventeen craft sailed to intercept an Allied convoy on 1 January 1945. Only two returned, the others falling victim to British destroyer escorts, to foul weather or to technical failures. For Hitler and Donitz the only consolation was the sinking of the British trawler *Hayburn Wyke*. Other fleets of Seehund craft were sent to raid British shipping but few reached their operational areas. The crews had a right to be fearful.

In March 1945, with Hitler desperate for success, twenty-nine Seehunds went forth for battle but the fighting was one-sided. Vigilant British frigates, MTBs and aircraft accounted for nearly all of them. True, a Seehund sank HMS *Puffin* but only because its over-confident skipper made the mistake of ramming the Seehund, whose torpedoes exploded into the vessel. Nevertheless, that month Seehunds sank three other ships, the largest being a vessel of nearly 3,000 tons.

Seehunds made 142 forays between January and May 1945, sinking in all nine ships totalling about 19,000 tons. The German's greatest success was probably in the number of Allied ships, 500 of them, and aircraft, 1,100, which were kept busy searching for Seehunds. This minor sapping of the Allied war effort had no effect on the outcome of the conflict. Even so, senior British naval officers, as well as intelligence units working specifically against the German Navy, said that had work started earlier on Seehund and had the crews been as well trained as their British and Italian counterparts, they could have inflicted great damage on the Allies. Hundreds of Seehunds running amok among the 5,000 ships of the Normandy invasion fleet would have wrought havoc. There was nothing wrong with the quality of the Seehund crews. They were brave raiders and they deserved better from their leaders.

21

THE MUSSOLINI SNATCH,
12 SEPTEMBER 1943

The dictator of Italy, Benito Mussolini – 'Il Duce' to his admiring people – was always going to be a problem for the Allies when he was captured or after the war ended. This posturing braggart had a limited intellect but he could rouse enormous crowds to fever pitch of adulation and, linking his nation's fortunes to alliance with Adolf Hitler, he tried to model himself on Hitler. He did not succeed in this and Hitler manipulated him at will so that Mussolini had no choice but to put his forces under the German generals' command. However, there remained a solid minority opposition to Mussolini and these partisans showed their hand on 26 July 1943. This was only sixteen days after the Allies had invaded Sicily, preparatory to landings in Italy itself.

Mussolini was seen in King Victor Emmanuel's palace in Rome on 26 July – and then he vanished. No intelligence service in the world, not that of the British or Americans, nor even of Nazi Germany, knew what had happened to him. In fact, he had been kidnapped by Italian forces who were intent on surrender to the Allies, when a suitable moment might present itself. Mussolini was to be a bargaining counter, a big one physically and a profoundly important one politically. His captors had taken him to a closed-down winter resort on Gran Sasso, a mountain range north of Rome. Their security was good because the only way up the mountain was by funicular railway and this was in their hands. If any contact with the outside world was needed then one of the band made it; nobody else was ever allowed up the mountain.

Adolf Hitler, however, wanted his partner back. The mission to retrieve Mussolini fell to Captain Otto Skorzeny of the Waffen SS and commander of a force modelled to an extent on the British Commandos. He had done well in creating and establishing his 'Special Troops'. Aged thirty-five, he might be a mere captain but his height of 6 feet 4 inches made him look imposing, as did the duelling scar from ear to chin. Also, he had a presence and a gift for self-publicity.

Skorzeny was enjoying a brief period of leave in Berlin when the mission to retrieve Mussolini was being formulated and by chance called his HQ at Friedenthal to ask his lieutenant and great comrade Karl Radl if anything was happening. Radl gasped with relief to hear Skorzeny's voice. 'Anything happening?!' he exclaimed. 'You are wanted at Hitler's HQ! A special plane is waiting for you at Tempelhof airport and it leaves at five this evening. You will be on it.'

Skorzeny was one of six officers summoned to see the Führer in his base at Rastenburg, in East Prussia. The other five were senior to Skorzeny and he came last as they were ushered into Hitler's presence. They stood in a row and Hitler came to each man in turn, looked into his eyes and asked for a brief résumé of his career. Then he stepped back to put a question to the group. 'Which of you knows Italy and what do you think of the Italians?'

The first five, playing safe, merely spoke about Germany's 'gallant allies', the Italians. Skorzeny could do much better. Looking Hitler full in the face, he said, 'Leader, I am Austrian.' Hitler, himself an Austrian, knew exactly what Skorzeny meant by this cryptic utterance. All patriotic Austrians resented the loss of the Alto Adige, Austria's most charming region, to the Italians at the end of the First World War. 'Captain Skorzeny,' Hitler said, 'you will stay behind. The rest may go.'

Once he was alone with Skorzeny, the Führer said, 'I have a mission of the highest importance for you.' It was some time before he completed the build-up, but he got there in the end. Mussolini had just been betrayed by King Victor Emmanuel and was under arrest. Italy was open to invasion and the conspirators intended that Italy should go over to them, with Mussolini as their great political gift. 'I cannot and will not leave Mussolini to his fate,' Hitler said. 'He must be rescued before the traitors can surrender him.' It was up to Skorzeny. Only five or six other people would know of his operation, including Himmler, chief of the SS, and General Kurt Student, the Airborne leader under whose command Skorzeny would act.

Skorzeny was unable to leave Berlin so Radl had the job of selecting the fifty very best men from Special Forces, including some who spoke Italian. Skorzeny ordered two machine-guns for every group of ten men and the same number of Schmeisser machine-pistols, 30 kg of plastic explosive, and grenades. Radl, an excellent staff officer, brought a greater variety of items than his chief had ordered, including some priests' outfits as a possible disguise.

Skorzeny and Student flew to Rome next day for talks with Field Marshal Kesselring, the German Commander-in-Chief in Italy, and others. The following day Radl arrived with the chosen fifty. Also sent from Germany was a strong detachment of paratroops in case their particular skills were necessary on this raid.

Shrewdly, Hitler tried to find out where Mussolini was being held and sent him some rare books for his sixtieth birthday on 29 July. He entrusted the books to a senior emissary, who approached Marshal Badoglio, successor to Mussolini, to ask if he might present them to Il Duce. Badoglio was not taken in. The books would be delivered to Mussolini, he said, but there could be no possibility of any German being allowed to know his whereabouts. Various agents suggested places where Mussolini could be held, including an island offshore from Naples. Skorzeny believed that his quarry was being moved from one place to another. On 18 August he was given an aircraft in which to fly over a certain fortified villa where Mussolini was reported to be. So as not to appear to be going straight for a particular place, Skorzeny ordered his pilot to fly over Corsica. Allied fighters shot his plane into the sea and Skorzeny and the aircrew had to be fished out of the water, Skorzeny with fractured ribs.

Remaining on duty, Skorzeny, disguised as an Italian naval rating, picked a conversation with a soldier standing as sentry to the fortified villa's gates. From him, Skorzeny heard that Mussolini had indeed been at the villa but he had left earlier that day in a white aeroplane with Red Cross markings. The next clue came from one of Skorzeny's agents who sent him a signal that had been intercepted at the Ministry of Home Affairs. It read: 'Security measures around Gran Sasso now complete'.

Skorzeny saw the implications at once. Security for what, at a non-functioning tourist resort nearly 10,000 feet up the Abruzzi mountains? A study of guidebooks told Skorzeny and Radl that a hotel named the Albergo-Rifugia was set on the plateau of Campo Imperatore at Gran Sasso. Flying at 15,000 feet, Skorzeny and Radl took photographs of the site but they hardly needed pictures to determine that the hotel could not be attacked by land. Even if he had an entire division of mountain troops under command, Skorzeny doubted the feasibility of a raid of such gigantic proportions.

Military gliders were being sent from France for Skorzeny's possible use. Having studied Witzig's report of his raid on Eben Emael, he had confidence in gliders. General Student, Skorzeny and the paratroop officer, Major Mors, decided on a plan. Mors would lead some paratroops to seize the lower funicular station in the valley. Skorzeny's Special Forces' fifty soldiers, together with fifty airborne men, such as medics and engineers should they be needed, would descend on the Gran Sasso plateau in twelve gliders. An important passenger was General Soleti, of the Carabinieri, who, it was hoped, would convince Il Duce of the German raiders' good intentions. Soleti, scared to death of the assignment, 'did a runner', but Radl located him in Rome and returned with him.

The raiders left their airfield early in the afternoon of 12 September and all seemed to be going well for the one-hour flight when Skorzeny discovered that his two lead groups were not part of the formation. Their towing aircraft had crashed into bomb holes created only minutes before by Allied attackers. The troops who were supposed to land first to give cover for the following groups had not even taken off. Then Skorzeny cut a hole in the canvas of his glider and looked through it to see that the plateau was not flat, as he had supposed, but as steep as a ski slope. There could be no turning back but Skorzeny was confident that enough raiders would survive the landing to seize Mussolini and prevent his guards from killing him, rather than have him fall into German hands.

However, the splendid pilot, Lieutenant von Berlepsch, managed to put Skorzeny's glider down, more or less in once piece, a mere 15 yards from the hotel. Moving swiftly, the raiders with Skorzeny came to a small room where an Italian soldier was seated at a radio. They smashed this set, disarmed and bound the Italian and raced along a terrace leading to the entrance. Skorzeny caught a glimpse of Mussolini's distinctive features at an upstairs window. No firing had taken place but Carabinieri were running from the hotel, only to be clubbed by the raiders' Schmeissers. Unless fired on they did not want to kill these Italians.

With his hand-picked men, Skorzeny raced upstairs to a room where Mussolini and two Italian officers were standing. They were disarmed and taken outside while Skorzeny and Mussolini had this brief exchange. As Skorzeny related it:

Mussolini: So everything's all right. I'm very grateful to you.
Skorzeny: Duce, the Führer has sent me. You are free.
Mussolini: I knew that my friend Adolf Hitler would not leave me in the lurch. I embrace my deliverer.

Radl and the support group had arrived and there was some firing before the colonel commanding the Carabinieri guard was brought in. He would not surrender, he protested; he needed time to consider. 'Of course you do,' Skorzeny agreed. 'You have sixty seconds.' He had his watch in one hand and pistol in the other. The disarmed colonel was allowed to leave the room but was back within the minute bearing a glass of wine, which he offered to Skorzeny as a token of surrender. From start to finish the raid had lasted only 12 minutes but several of Skorzeny's men had been killed in glider crashes. He now had to decide on the best way out for Mussolini but his deliberations did not take long for during a journey of more than 100 miles by road anything might happen to Il Duce. He would have to be flown out.

There now took place perhaps the most remarkable part of the exploit. By prior arrangement, two Fieseler Storch scout planes had landed on an airstrip in the valley, but only one was now serviceable and the pilot, Captain Gerlach, flew onto the plateau. Skorzeny did not want to leave Mussolini to travel alone on his escape flight. He was responsible to the Führer for this man's safety and he told Gerlach that he would accompany Il Duce in the plane. Gerlach was horrified. The Storch was a two-person aircraft, he pointed out, and between them Mussolini and Skorzeny were in weight three persons, in effect making the party four in all. Did Skorzeny really believe that he could fly off the plateau with such a load? Yes, Skorzeny said, he had implicit trust in Gerlach, as would the Führer himself.

He and Mussolini squeezed into the rear seat and the unhappy Gerlach took his place. Il Duce, a genuinely accomplished flier himself, was equally unhappy. Radl organised the raiders to hold the Storch back along its wings while Gerlach roared the engine to its capacity rpm. On a signal from Gerlach, they let go and the Storch raced forward. But it never did get airborne in the conventional sense – it simply fell over the plateau's edge. Gerlach controlled the fall, steadied the Storch and an hour later it reached Rome, where Mussolini was transferred to a Heinkel bomber which first took him to Vienna and then on to Hitler, awaiting his friend at Rastenberg.

Hitler promoted Skorzeny to major at once and that night he was presented with the coveted Knight's Cross of the Iron Cross. It was said to have been the first time that a Knight's Cross was awarded on the day of the exploit which it recognised. Skorzeny had certainly done well, though the raid's success was as much to the credit of the para Major Mors and to General Student himself.

Skorzeny carried out other daring raids and while they were perhaps more important than the rescue of Mussolini they have received relatively little attention. In September 1944 he prevented Hungary from concluding a separate peace with the USSR and rescued a million encircled German troops; he achieved this staggering coup by kidnapping the son of the Hungarian regent and at the same time militarily occupying the Citadel of Budapest, thus giving his political masters great leverage in negotiations.

During Hitler's Ardennes offensive – 'the Battle of the Bulge' – Skorzeny organised 'American Brigades' of disguised Germans to cause havoc behind Allied lines. The fear which infected Allied leaders and soldiers created the havoc – Skorzeny and his men needed to do very little beyond sowing the seeds of panic among the American divisions and their generals.

After the war ended Skorzeny, now a major-general, was declared to be 'the most dangerous man in Europe'. One of his several alleged war crimes was that he had fought in enemy uniform, contrary to international law. Many people believed that he might be hanged and but for a masterstroke by his American lawyer execution might well have been the verdict. As a defence witness, the lawyer called one of Britain's greatest Resistance war heroes, Wing Commander Forrest Yeo-Thomas, code-named 'the White Rabbit'. Yeo-Thomas volunteered the information that the British often wore enemy uniform in action. The case against Skorzeny collapsed but the Allied prosecution authorities were determined to nail him one way or another and he remained a prisoner for years.

This flamboyant man was a master raider along bold individualistic lines. Like David Stirling, he was impatient with obstructive authority, which at times plagued him, and he broke through it by dubious means. Again like Stirling, he requisitioned arms, transport, and supplies by invoking names that could not be denied – that of Hitler himself, if necessary. At times he simply took from dumps whatever he needed – rather like Stirling when he ordered his men to steal a camp for themselves. Unlike Stirling, he was duly recognised by his superiors, so that he won promotions, decorations and fame.

The password chosen by Skorzeny and Radl for the Mussolini caper was *Immer mit der Ruhe!* – meaning, 'Take it easy!'. The choice of this phrase says much about Skorzeny's personality. While he could be tempestuous, he always did 'take it easy'. As an attitude, this is one of the most desirable qualities for raiders.

22

THE RAID ON AMIENS
PRISON, 18 FEBRUARY 1944

OPERATION JERICHO

Air raids of the type which flattened cities and destroyed industrial complexes cannot form part of this book because they do not fall within my definition of a raid, in the way that SAS, LRDG, SBS and earlier Commando forays were raids. I doubt if any writer on war would consider as raids, say, the bombing of Berlin, Coventry, London, Tokyo and Dresden. These were military assaults on a massive scale but in journalese 'raid' was such an easy label. Many thousands of such 'raids' took place during the war, so many that even official records give many of them only a line or two.

There were exceptions. The 'Dambusters' attack on the Ruhr reservoirs was a raid of epic proportions. Pinpoint air raids were also made against certain Gestapo headquarters and on numerous occasions the pilots of light bombers were sent to destroy bridges which the enemy were using to bring up reinforcements and so threaten Allied movements. One brief operation by the RAF ranks as one of the most remarkable military actions in history and it was unique in that its purpose was to save lives, not to take them. It was an epic event on a level with the Dambusters' operations and it called for the same great skill, daring and sheer bravery. It has caused controversy ever since.

The raid took place on 18 February 1944 and the RAF's target was Amiens prison, in France. The reasons for which the attack was mounted, the quality and degree of planning required and the level of courage displayed – by the French Resistance as well as by the aircrews concerned – make an intriguing story.

Hitler and his generals knew that at some time during 1944 the Allies would launch an invasion against the occupying German forces in France and the order went out to the Gestapo and the Abwehr, the field security police, to smash the French Resistance cells so that they could not give the invaders any effective support.

Particular attention was given to the area of north-west France where the Nazis judged the invasion would take place. Following this instruction, thousands of French men and women were arrested on suspicion of involvement with the Resistance and the French civil prisons, taken over by the Germans, filled up. Many of those incarcerated were undoubtedly Resistance people but the Nazis seized anybody who just might be considered dangerous. In this they were assisted by numerous French collaborators and traitors and the pro-German police, the Milice. Overcrowding was dreadful, with up to eight people in a cell intended for three. Prisoners took turns to sleep, since there was not enough space for all to lie down at once. The cold was intense and the food vile.

The German counter-Resistance forces were more successful than they knew. They had broken up, captured or wiped out several valuable networks, a grim fact understood by British and Free French intelligence services in London. It could not have happened at a worse time for the British because the Nazis were building up their sites for launching their V1 and V2 rocket missiles on Britain and the British leaders had counted on the Resistance for information.

During January 1944 it was known in London that more than 100 important Resistance people were awaiting trials in Amiens prison. They were trials in name only; nobody was ever found innocent and acquitted and the guilty verdict was a mere formality. After the trials came executions, usually by firing squad.The next batch of judicial murders was to take place in mid-February. Senior Resistance men in France sent a message to London appealing to the British military to bomb the walls of Amiens prison to enable prisoners to escape.

At first, the Resistance chiefs did not specify what type of 'bombing' they meant, but it did not take a feasibility study to show that the prison could not be attacked from the ground without an enormous number of casualties. There was not a chance in a million that even a Commando attack and a paratroop strike together could do enough damage to the prison quickly enough to permit a mass escape. Within minutes of such an operation beginning the Germans could call in entire divisions of troops, including heavy armour. An air raid using 500 lb bombs was the only conceivable form of attack on a prison where the walls were 20 feet high and 3 feet thick. The problem was handed to Air Chief Marshal Sir Arthur Coningham who discussed it with Air Vice Marshal Basil Embry, who was to be the principal planner. Nobody liked the idea. It would be dangerous for the RAF crews involved and infinitely more hazardous for the Amiens prisoners. The bombs could so easily destroy the entire structure of the prison and kill hundreds of innocent people. That some Germans would die with them was insufficient compensation.

The Resistance leaders were serious in their requests and they sent much information about the prison and the way it functioned. They had even obtained a copy of the original builders' plans for the place. They provided a layout of the buildings within the massive walls and numerous fine drawings added to a description. Included in the information were details of where the German guards were housed, together with their eating and sleeping patterns. RAF reconnaissance fliers photographed the prison from heights safe enough not to have the Nazis suspect that pictures of the prison were being taken.

The decision to make the raid, using the fast, highly manoeuvrable Mosquito bombers, was reached only after agonising discussions. Basil Embry was the best possible choice as master planner. He owed his life to the Resistance. In 1940, after being shot down over France, he was captured by the Germans but killed his guards and escaped. The Resistance then passed him down an escape line and eventually he reached Britain. He had every reason to help the Resistance now, just as he had powerful reasons for not wishing to injure these brave people. He often had to remind himself, and others, that the raid was being made at their request.

As well as Resistance workers and common civil criminals, five espionage agents were in solitary in Amiens prison – three British spies, one American and one Belgian. Three other Americans, captured in civilian clothing, were in cells as suspected spies, though they claimed to be shot-down airmen. All these men had to be given a chance to get out. For the Allies, as well as for the Germans, the most important prisoner was Raymond Vivant, the Under-Prefect of the city of Abbéville, near Amiens. He knew so much about the Resistance groups, escape lines and German dispositions in north-west France that the Germans could not afford to lose him. He was still alive because the Nazis still did not know all of Vivant's knowledge and because they had yet to break him. Because of that same knowledge the Allies and the Resistance chiefs wanted him free but if this were not possible they wanted him dead so that he could not talk.

Basil Embry wanted to lead the aircraft making the attack himself but this was wisely vetoed by his superiors. He was too valuable to be risked and were he to crash and be captured he would be a great prize for the Nazis. They would realise that such a high-ranking officer possessed military secrets and they would go to any lengths of interrogation by the Gestapo sadists to draw these secrets from him. In any case, they probably knew he was the Englishman who had killed his German guards.

Command of the attacking force went to Group Captain Charles Pickard DSO and two Bars, DFC. One of the most able of all RAF leaders in the air, Pickard was greatly respected by his men, who knew him familiarly as 'Old Pick', 'the Pied Piper of Hunsdon' – the airfield from which he flew. Pickard was famous to millions of people around the world as the captain of Wellington bomber 'F for Freddie', the star of a propaganda film about a bomber raid, called *Target for Tonight*. Under Pickard were three Wing-Commanders, I.S. 'Black' Smith DFC and Bar of the New Zealand 487 Squadron, 'Daddy' Dale of the RAF's 21 Squadron and R.C. 'Bob' Iredale of the Australian 464 Squadron. Pickard's navigator was Flight Lieutenant J.A. Broadley DSO, DFC, DFM.

The first code name for the mission was Operation Renovate but Embry changed it to Operation Jericho, a singularly appropriate label for an action designed to bring down walls. Very few people knew of Embry's planning, not even the aircrew. An unusual senior officer, Embry was impatient with bureaucrats and against Air Ministry opposition he had earlier established his own planning and modelling committee for just such an operation as Jericho. His hand-picked experts went about their work without any reference to any 'higher authority', other than Embry himself. Under the guidance of Intelligence Staff Officer Squadron Leader Turner-Lord, a former architect, a group of specialists made a large-scale plaster model of the prison and the buildings within the walls. Others made relief models of the area around Amiens, which the pilots and their navigators would study until instinctively they knew the best lines of approach to the targets and the best way out. They were lucky in that the large building was by the side of the dead-straight Amiens–Albert road, which acted as an aiming strip.

At their first briefing the crews were startled by what they heard but after glances all round they took the news calmly. In essence, crews of 487 Squadron would be the first wave, their task to breach the walls on the north and east sides. A second wave was to open up either end of the prison itself and destroy the garrison's quarters. A third wave was to stand by, ready to be called in by Pickard if the first two had not done sufficient damage to enable a mass escape to take place. Squadron Leader Tony Wickham, flying DZ414 of the Film Production Unit, would film the raid while escort and cover for the Mosquitoes was provided by Typhoons of 198 Squadron.

The most daunting information given to the crews was that the attack run had to go in at an altitude of about 20 feet, enough to clear the wall. In the event, some pilots came down to 10 feet. For fliers other than the most highly skilled these were suicidal 'altitudes'.

Embry told his crew, 'There has been much debate as to whether this raid is feasible. It can certainly only be carried out by low-level Mosquitoes. Inside that prison are more than one hundred patriots who have been condemned to death and are to be shot at any moment for assisting the Allies and there are hundreds more men and women expecting similar fates, some for helping airmen like us shot down in France.' Then Pickard emphasised the difficulties of the attack:

You've got to be right down on the deck and if we aren't damned careful our bombs are going to bounce right over the walls, land inside the prison grounds and blow everyone to smithereens. We must cut that risk to a minimum. You have to be below – below – the height of the wall when you let go; down to 10 feet if possible. There are no obstructions on the run up, so you should be able to make it. Timing is essential if we are to avoid blowing one another up with our bombs. We are dropping 11-second delay bombs. You can see from the model that the prison itself is in the form of a cross. At its east and west ends are smaller buildings which are said to be quarters for the prison staff. The second six aircraft are going to prang [demolish] those quarters. I don't suppose all the Nazis will be inside at the one time but we're sure to get some of them and it will add to the general confusion, giving prisoners a better break.

The crews were uneasy when told that the Resistance would be notified in advance of the precise time of the attack. They suspected what British Intelligence knew very well, that the Resistance had its traitors and informers so there was a risk that the raid would be betrayed to the Nazis. However, Embry and Pickard pointed out, it was imperative that the political prisoners in the prison should know of the impending attack so that they could be ready to run. It was also important for Resistance workers on the outside to be informed so that they could organise safe houses, transport and escape routes.

About a hundred local people outside the prison and sixteen inside, mostly from the Sosie network, were told that the raid would take place at noon on any day in a particular period. Those outside became very busy, though cautiously so. The bicycle repairer, a key network member, placed a supply of bikes at houses and shops close to the prison. People who owned a car or truck were recruited and organised – providing that they could give enemy patrols a valid reason for being on the roads.

Nineteen raiders took off from Hunsdon in a swirling snowstorm, which offered a certain amount of cover, but the snow eased as they crossed into France. Everywhere was one great sheet of white. All crews knew that enemy radar would pick them up, despite their low altitude. The target was clear and the approach run perfect and, as urged by Pickard at the final briefing, 'Black' Smith released his bombs at a height of 10 feet, that is, from half the altitude of his target. He and the two pilots flying with him, had pulled their sticks back hard and fast and the Mosquitoes came up sharply as the bombs left them. Their momentum carried them angle-up for long enough to clear the wall and then sped towards their targets. The Mosquitoes came up abruptly, roaring low across the prison buildings. Then, banking so sharply that they vibrated violently, the aircraft zoomed away practically at street level. Many terrified pedestrians flung themselves down or sheltered in doorways. Some bombs went clean through or over the east wall, wobbled between buildings as though on loose guidewires, and smashed into the west wall. The time-fuses had been accurately set and 11 seconds later the massive 500-pounders blew up, creating great destruction.

One bomb, dropped from Bob Iredale's Mosquito, made a direct hit on the guards' quarters and another smashed the main building. Amazed, some pilots saw a few bombs slithering wildly across the snow outside the prison, demolishing houses and fences before their 300 mph momentum ran out. Then up they went. Squadron Leader McRichie's bombs hit the guardhouse and walls. By now Amiens prison was blotted out by smoke and dust. As planned, Pickard, who had been flying with Iredale's squadron, pulled away and climbed to 500 feet to observe results. Most crews saw ant-size figures running from the gaps in the walls and onto the snow-covered field. Some fell, picked themselves up and continued their frenzied flight. Pickard and Broadley had a grandstand view of the exodus. Pickard made up his mind quickly – no further damage was necessary and by radio he snapped the words that would send 'Daddy' Dale and his section back to England: 'RED DADDY – RED DADDY.' Some records give this as DADDY – DADDY – RED – RED – RED.

Pickard and Broadley were the last airmen to leave the scene. It is possible that their Mosquito was damaged by ground fire before two Focke Wulf 190s attacked, shooting off the plane's tail. The wreckage fell near Montigny, with the two fliers probably already dead. Some civilians reached the scene before the Germans and a girl cut off Pickard's rank insignia, wings and medal ribbons, which were later returned to his widow. Somebody else scooped up navigator Broadley's maps and papers and rushed away to bury them. They too were retrieved and handed to a British officer when the Nazis pulled out.

Broadley and Pickard were buried next day close to the prison. Of the nineteen Mosquitoes which had set off on the raid, three were damaged, one beyond repair, and two failed to return. An escorting Typhoon was shot down near Amiens, another crashed in the Channel and a third made a forced landing. The losses, though grievous, could have been much greater.

The escapees ran in all directions, anything to get away from the vicinity of the prison, though a few found safe hiding places nearby until they could change out of their distinctive prison garb. Some members of the ad hoc 'reception committee' had collected large quantities of clothing, ready for just such an eventuality. Five escapees got away by hiding in a baker's van. A soldier stopped the driver and demanded to know what he had inside. Coolly the driver said, 'Six fresh-baked escaped prisoners – what do you think I've got!' The soldier impatiently waved him on. Other men hid in one or other of the Amiens brothels. Many took to the snow-covered fields and without heavy over-clothing and sound footwear they suffered greatly from the cold.

The Germans and their collaborators mounted a huge manhunt. On orders from Berlin, probably from Hitler himself, armoured cars, tanks, car and motorcycle patrols and truckloads of troops descended on the entire Amiens district. Virtually nothing escaped the attention of the infuriated Germans. They sprayed haystacks with machine-gun fire and stuck a bayonet into anything that might just conceal a prisoner. Rather than go to the bother of a minute search, the raiders set fire to some buildings. The hunt became so intense that nowhere in the entire Somme *département* – as big as a British county – was safe. The established escape lines, which normally dealt with shot-down airmen on the run and the occasional soldier escaper or evader, became extremely active in passing on Amiens prison escapees.

In their methodical way, the German authorities in Amiens drew up a list of 'categories' after the raid, but omitted the number of Germans killed or wounded. Resistance estimates were that 20 had been killed and 70 wounded but these figures could never be verified. Before the bombing, according to the Germans, the prison held 448 male prisoners and 72 females. The German staff comprised 180 men and women. Therefore, on Nazi reckoning, 700 people were inside the prison at midday on 18 February – that is, if the Nazi figures can be believed:

AFTER THE BOMBING

Recaptured	182
Held in a women's transit centre, in rue de la Republique	26
Kept in the German quarters, Dury	8
Men in the Citadelle	48
Held by the Chief Guard	20
Killed during the raid	87
Wounded and in hospital	74

This made a total of 445, leaving 255 'missing'. Not all were political prisoners. Through the battered walls, with the Resistance people, had scurried criminals of all types, from pickpockets and thieves to men under investigation for civilian murder.

The German figures are suspect. A gendarme who was himself a prisoner in the gaol, accused of helping 'terrorists' – that is, Resistance workers – produced a different list. While he confirmed the figure of 180 in the German quarters, his analysis listed 640 detainees. But even this figure might not be correct. A police chief in the Somme *département* reported that the prison held 840 civil or political prisoners at the time of the raid with another 180 locked up by order of the Germans. The distinction between those 'locked up' and those actually imprisoned was never clear but obviously the number of people who escaped was somewhat greater than the official figure. The supposition was that the Germans were trying to fog the true picture.

They certainly fogged the fate of Raymond Vivant, the Abbéville Under-Prefect so desperately wanted by Allied Intelligence. The Nazis' hunt for this one man was rigorous but largely with the help of his intelligent and daring wife he evaded the dragnet. The Germans claimed that he was killed in the raid and went through the formality of 'burying' him. After the war when the coffin was exhumed Vivant's coat and hat were found in it. Vivant joined General de Gaulle and served in his Ministry of the Interior.

It is said that the raid and the consequent escape gave the Resistance encouragement that was desperately needed at the time in the face of German successes in curtailing its activities. When the Normandy invasion came in June the Resistance was ready to do its part. Escaped Resistance workers and other 'politicals' fingered about seventy Gestapo agents and their French collaborators. Their names and whereabouts were known and several were murdered. German counter-intelligence chiefs were forced to transfer their Gestapo thugs away from the region and to find somewhere safe for their French helpers. The entire system of oppression in the region fell apart. There was never a safe haven for the collaborators; sooner or later they paid the price for their treason, sometimes officially after a trial, more commonly 'unofficially'. They had always assumed that no witnesses would survive to confront them with their crimes. Operation Jericho put that right.

Was it morally right to bomb the prison knowing that many French people and a sprinkling of other nationalities would certainly die? Ethics became a matter of balance. It is known, from captured German records, that twelve prisoners were to be shot on 19 February and other executions would have gone on for months. Nobody knows what the total would have been before the occupation ended. The air crews found some comfort in the assurance from the Resistance that the risks had to be taken and in the messages of support later given to the RAF for the decision to bomb.

Commenting on the raid, an Australian navigator said, 'This was the sort of operation that gave you the feeling that if you did nothing else in the war, you had done something.'

The raid was the right decision at the time, otherwise the British, to whom the request came, would have been surrendering the moral high ground. Here was an opportunity to put principles into practice, while deeply regretting some of the consequences. The blame for the deaths of innocent people must be laid at the feet of Hitler and his Nazis.

I think it is significant that local people near the prison took away the markings left by the Germans and put crosses over the graves of Pickard and Broadley and later they inscribed the airmen's names, rank and decorations. On the grave they placed flowers, which were never allowed to remain there withered. Fresh flowers just 'appeared'. Perhaps these acts by people close to the site of the raid – and perhaps even involved in the escape in one way or another – is sufficient response to those who condemned the raid as 'immoral', 'unethical' and 'murderous'. Having lived through the period of Nazi barbarism which seemed to have no end, the people of Amiens most affected by the Embry–Pickard raid understood that the RAF crews did what they had to do.

BIBLIOGRAPHY

This is very much a select bibliography. Many books have been published about the battles in which Commandos fought, about naval, air and army operations involving 'special forces' and about the various units which might be called commandos, raiders, independent companies, and other outfits. A list of all of them would require a book in itself. Those I mention here are the most prominent in their field, though some cover a much longer period than 1939–45. Some titles are also mentioned in the text. Chief among these are: Seymour's *British Special Forces*, Lodwick's *The Filibusters* (also known as *Raiders from the Sea*) and Kemp's *The SAS at War 1941–1945*. Macksey's *Commando Strike* is also a major book in this field. Other information on sources is given in the text. Details about the remarkable exploits of Rex Blow DSO and Jock McLaren came to me from Blow himself and reports in Australian Archives. Unless otherwise stated, the books referred to in the Bibliography were published in London.

Admiralty & Combined Operations records in the Public Record Office

Bekker, C., *K-Men: The Story of German Frogmen and Midget Submarines*, William Kimber, 1955

Borghese, J.V., *Sea Devils*, Andrew Melrose, 1954

Buckley, Christopher, *Norway, The Commandos, Dieppe*, HMSO, 1977

Challenor, Tanky with Draper, Alfred, *SAS and the Met*, Leo Cooper, 1990

Connell, Brian, *The Return of the Tiger*, Evans Brothers, 1960

Courtney, G.B., *SBS in World War Two*, Robert Hale, 1983

Cowles, Virginia, *The Phantom Major*, Collins, 1958

Durnford-Slater, John, *Commando*, William Kimber, 1953

Fishman, Jack, *And the Walls Came Tumbling Down*, Souvenir Press, 1982

Foley, Charles, *Commando Extraordinary – Otto Skorzeny*, Longman, 1954

Frere-Cook, O., *The Attacks on the Tirpitz*, Ian Allan, 1973

Garrett, Richard, *The Raiders: The World's Elite Strike Forces*, David & Charles, 1980

Hampshire, Cecil, *The Beachhead Commandos*, William Kimber, 1983

Harrison, D.I., *These Men are Dangerous: The Special Air Service at War*, Cassell, 1957

Kemp, Anthony, *The SAS at War 1941–1945*, John Murray, 1991; Signet 1993

Kemp, Paul, *Underwater Warriors*, Arms & Armour, 1996

Kennedy Shaw, W.B., *Long Range Desert Group*, Collins, 1945

Lassen, Suzanne, *Anders Lassen VC*, Muller, 1965

Lewis, Jon E., *Giant Book of SAS*, The Book Company, 1995

Lloyd-Owen, David, *The Desert My Dwelling Place*, Cassell, 1957

——, *Providence Their Guide*, Harrap, 1980

Lodwick, John, *The Filibusters*, Methuen, 1947, republished as *Raiders from the Sea* and produced by Greenhill in 1990

Macksey, Kenneth, *Commando Strike: The Story of Amphibious Raiding in World War II*, Guild Publishing 1985

McKie, Ronald, *The Heroes*, Sydney, Angus & Robertson, 1960

Millar, George, *The Bruneval Raid*, 1974

O'Neil, Richard, *Suicide Squads*, Salamander/Lansdowne, 1981

Phillips, C.E. Lucas, *Cockleshell Heroes*, Heinemann, 1956

Robertson, Terrence, *Dieppe: The Shame and the Glory*, Hutchinson, 1963

Ryder, Commander R.E.D., *The Attack on St Nazaire*, John Murray, 1947

Seymour, William, *British Special Forces*, Sidgwick & Jackson, 1985

Skorzeny, Otto, *Special Mission*, Futura, 1974

St George Saunders, Hilary, *The Green Beret: The Story of the Commandos*, Michael Joseph, 1949

Strutton, Bill and Pearson, Michael, *The Secret Invaders*, Hodder & Stoughton, 1958

Warner, Philip, *The SBS*, Sphere, 1983

Warren, C.E.T. and Benson, James, *Midget Raiders*, NY, William Sloan, 1954

Young, Peter, *Commando*, NY, Ballantine Books, 1969

INDEX